AI
HORIZONS

AI
HORIZONS

Shaping a Better Future Through
Responsible Innovation and Human Collaboration

ENAMUL HAQUE

MERCURY LEARNING AND INFORMATION
Boston, Massachusetts

Publisher: David Pallai
MERCURY LEARNING AND INFORMATION
121 High Street, 3rd Floor
Boston, MA 02110
info@merclearning.com
www.merclearning.com
800-232-0223

E. Haque. *AI Horizons: Shaping a Better Future Through Responsible Innovation and Human Collaboration.*
ISBN: 978-1-50152-169-0

The publisher recognizes and respects all marks used by companies, manufacturers, and developers as a means to distinguish their products. All brand names and product names mentioned in this book are trademarks or service marks of their respective companies. Any omission or misuse (of any kind) of service marks or trademarks, etc. is not an attempt to infringe on the property of others.

Library of Congress Control Number: 2024938114

242526321   This book is printed on acid-free paper in the United States of America.

Our titles are available for adoption, license, or bulk purchase by institutions, corporations, etc. For additional information, please contact the Customer Service Dept. at 800-232-0223(toll free).

All of our titles are available in digital format at academiccourseware.com and other digital vendors. The sole obligation of MERCURY LEARNING AND INFORMATION to the purchaser is to replace the files based on defective materials or faulty workmanship, but not based on the operation or functionality of the product.

TO

Professor Paolo Zanella, a luminary of wisdom, a source of motivation, and my inspiration to study computer science at the University of Geneva. His unwavering dedication to the field of computer science and the education of his students left an indelible mark on my own journey. His guidance shaped my understanding of computer science and instilled in me a passion for sharing knowledge with others. With immense gratitude and admiration, I dedicate this book to him. May his legacy continue to inspire future generations.

CONTENTS

CHAPTER 2: AI APPLICATIONS AND REAL-WORLD CASE STUDIES **77**

PREFACE

In a not-so-distant future, a child will ask their parent, "What did you do to ensure artificial intelligence benefited humanity?" This is a question we all must be prepared to answer. As we stand on the brink of a technological revolution that promises to redefine the fabric of our existence, we must confront the myriad ethical, social, and global challenges that artificial intelligence (AI) presents. *AI Horizons: Shaping a Better Future Through Responsible Innovation and Human Collaboration* explores these challenges and provides a roadmap for navigating a future intertwined with AI.

As you read these words, algorithms make decisions that impact your life. They determine the news you see, the products you buy, and the job opportunities available. Yet, these algorithms do not possess consciousness. They do not feel joy, anger, or fear. They are compelling tools capable of transforming industries, addressing global challenges, and unlocking new possibilities. However, without careful consideration and responsible governance, these tools could also exacerbate inequalities, undermine democracies, and concentrate power in the hands of a few.

Just as cows were domesticated thousands of years ago, humans now stand on the verge of being domesticated by algorithms. These algorithms, empowered by AI, can manipulate our emotions, decisions, and, ultimately, our lives. Yet, unlike the cows, we possess the power to shape our future and decide whether to become passive subjects of algorithms or architects of a better destiny.

As we grapple with the rapidly evolving landscape of artificial intelligence, we must also confront the harsh reality of a digital divide that threatens to leave vast segments of the global population behind. Bridging this divide requires technological innovation and a commitment to inclusivity, fairness, and social justice. It requires recognizing that AI is not just a tool for economic growth but also a catalyst for social change and global solidarity. As we strive to build a better future through AI, we must address the pressing global challenges of poverty, inequality, and environmental degradation. More is needed to develop intelligent machines. We must also work toward a more just, sustainable, and equitable world.

Human collaboration and oversight are paramount in a world increasingly dominated by AI. We must develop robust accountability, transparency, and ethical decision-making frameworks as we delegate more decisions to machines. We must foster a culture of collaboration that brings together diverse stakeholders, including technologists, policymakers, and civil society, to shape the future of artificial intelligence. Only through collective action and a shared commitment to human-centric values can we ensure that artificial intelligence serves the common good and contributes to a brighter future for all.

Consciousness, Intelligence, and Responsibility

Science fiction often conflates intelligence with consciousness, envisioning a future where AI gains consciousness and decides the fate of humanity. However, intelligence, the ability to solve problems, and consciousness, the ability to feel things, are distinct concepts. While we cannot rule out AI's possibility to develop consciousness, it is not a prerequisite for high intelligence. Just as airplanes fly without feathers, AI can solve problems without feelings. The challenge, then, is not to prevent a robot rebellion, but to ensure that the immense power of AI is harnessed for the betterment of all rather than the detriment of many.

Navigating the Future

In the following chapters, we will examine the ethical considerations, social consequences, and global implications of AI. Through real-world case studies, we will explore how AI transforms industries, from health care and education to transportation and finance. We will investigate the adoption of AI in emerging markets and its role in addressing global challenges such as poverty, climate change, and sustainable development. We will also delve into the necessity of human-AI collaboration, the importance of human-centric design, and the need for comprehensive governance and policy frameworks.

A Call to Action

This book is more than a primer on AI: it is an invitation to actively participate in shaping its impact. It is a call to arm yourself with knowledge, engage in crucial conversations, and advocate for policies that center human well-being alongside innovation. Imagine a future where AI is not something that merely happens to us, but a force we collectively and intentionally guide. Consider these actions:

1. *Educate Yourself*: Dig deeper than headlines. Seek out varied perspectives on AI, such as the potential benefits, the ethical risks, and the global implications.

2. *Join the Dialogue*: Talk to friends, family, and colleagues about your AI concerns and hopes. Participate in community discussions, attend workshops, or become involved with ethical AI initiatives.

3. *Demand Accountability*: Support businesses and policymakers who prioritize transparency, fairness, and bias mitigation in AI systems. Be vocal about addressing the digital divide and ensuring AI benefits extend to all.

4. *Foster a Human-AI Alliance*: Explore how ethical AI use can enhance your skills and field. Look for ways to combine human expertise with AI's power to solve problems in your workplace, community, or your daily life.

The future of AI is being written every day, and your voice matters. By making informed choices, demanding responsible practices, and embracing the potential of human-AI collaboration, you help ensure that the child asking that pivotal question will live in a world where the answer is one we can all be proud of.

Book Overview

Chapter 1: *Focusing on AI Ethics and Social Impact*

This chapter explores AI's ethical considerations and social consequences, examining privacy, fairness, accountability, transparency, and the digital divide. It also studies the role of AI in changing the future of labor, education, health care, and other industries. As AI becomes increasingly intertwined with our daily lives, it is crucial to understand and address its ethical and social implications.

Chapter 2: *AI Applications and Case Studies*

Presenting realistic examples and real-world case studies of AI adoption across diverse industries, this chapter helps readers better comprehend how AI is transforming businesses and society. AI is revolutionizing our lives and work, from health care and education to transportation and finance.

Chapter 3: *AI in Emerging Markets*

This chapter investigates how AI is accepted and adopted in emerging economies, highlighting the problems and opportunities these places bring. It also studies how AI may address significant global concerns such as poverty, climate change, and sustainable development.

Chapter 4: *The Human-AI Partnership*

Examining the necessity of collaboration between humans and AI, this chapter stresses the need for a symbiotic relationship where AI augments human intelligence and creativity rather than replacing it. This perspective underlines the significance of human-centric design and responsible AI development.

Chapter 5: *AI Governance and Policy*

Analyzing the role of governments, regulatory organizations, and international groups in defining the future of AI, this chapter examines the necessity of developing adequate legal and ethical frameworks to ensure the proper development and use of AI technologies.

Chapter 6: *Public Perception, Acceptance, and Literacy of AI*

This chapter examines how the public perceives and accepts AI, emphasizing the importance of AI literacy. It explores fostering critical thinking and informed optimism about AI technologies. It discusses promoting AI literacy to empower individuals to navigate an AI-driven world. It highlights the role of multi-stakeholder collaboration in ensuring diverse voices shape AI's ethical and societal implications. Understanding public perception, promoting acceptance, and enhancing AI literacy is essential for a well-informed society that leverages AI's benefits while mitigating risks.

Chapter 7: *The AI Risk Factors and How to Mitigate Them*

This chapter discusses the importance of education, retraining, and social safety nets in the face of job losses brought on by technology. It explores the need for multi-stakeholder initiatives to share knowledge, resources, and best practices for AI development through collaboration across governments, businesses, and academic institutions.

Chapter 8: *Generative AI*

Providing a comprehensive overview of generative AI, its applications, and its impact on various industries, this chapter ensures that the book covers a broader spectrum of AI technologies and their implications.

E. Haque
August 2024

*I*NTRODUCTION

We are in a time of profound change, which parallels the advent of the Internet in its potential to revolutionize how we live, work, and think. This transformation arises from artificial intelligence (AI), a technology that challenges our understanding of intelligence and consciousness. In this book, we will delve deep into the world of AI, examining its origins, current applications, future potential, and the ethical dilemmas it presents.

Understanding Artificial Intelligence

At its core, AI is about crafting computer systems capable of handling tasks that, until now, have typically demanded human intelligence. This includes a range of activities such as recognizing and interpreting visual data, understanding and responding to speech, making decisions based on complex information, and translating languages. Central to AI is machines' ability to learn from experience, adjust to new inputs, and perform human-like tasks. This learning aspect, a fundamental feature of AI, propels us into the realm of machine learning, a critical subset of AI that focuses on developing systems that can grow and change independently.

Machine Learning

Machine learning (ML) represents a specialized branch within the broader field of artificial intelligence, focusing primarily on crafting computer programs capable of learning and evolving through experience. At its heart, ML revolves around developing algorithms designed to sift through and analyze extensive datasets, discerning patterns within. This process empowers these systems to make informed predictions or decisions based on their analyses.

Crucially, ML algorithms are broadly classified into three categories: supervised learning, unsupervised learning, and reinforcement learning. Each type represents a unique method for training these algorithms, each with its distinct set of uses and inherent limitations. Supervised learning involves training the model on a labeled dataset, teaching it to recognize patterns and make predictions. Unsupervised learning, in contrast, deals with unlabeled data, allowing the algorithm to identify structures and patterns on its own. Reinforcement learning is a dynamic process where algorithms learn to make decisions by performing actions and observing the results or feedback from those actions. Each of these learning types unlocks different capabilities and applications in machine learning.

Deep Learning

Deep learning (DL), a specialized area within machine learning, uses what we call "neural networks" that have many layers, making them "deep." Imagine trying to mimic how our brains work; with all their complex connections, that is what these networks attempt to do. They learn from enormous amounts of data.

A neural network with just one layer can do some basic guesswork, much making a rough sketch. As you add more layers, the picture gets clearer and more detailed. These extra layers help the machine be more accurate and identify complicated patterns. It is like having a team where each member looks at a problem from a different angle and figures it out better.

Discriminative and Generative Models

In machine learning, we often distinguish between two main types of models: discriminative and generative. Think of *discriminative* models as the sort that specialize in telling things apart. They differentiate between several types of data. For instance, discriminative models determine whether an email you received is just another spam message or something more substantial.

Generative models are "creative." They create something new that resembles a given set of data. These models create new images, craft melodies, or write text. They use existing examples to generate fresh, similar instances.

Generative AI (GenAI)

Generative AI, controlled by large language models, is like a digital artist capable of producing original pieces, whether it is music, art, or blocks of text. These models are not just replicating; they are inventing, using their training to fabricate new data that mirrors the examples they have learned from. An example in this field is ChatGPT from OpenAI, renowned for its ability to generate strikingly human-like text based on the prompts it receives. We will examine the mechanics and implications of generative AI, including a simpler breakdown of how it functions, in Chapter 8 of this book.

Large Language Models and Generative Language Models

Large Language Models (LLMs) are neural networks fed with an enormous amount of text data, and they are remarkable for their language processing ability. LLMs are trained to digest this wealth of information and then generate text that appears human-like. You have likely encountered them in the form of chatbots or content creation tools.

Generative Language Models (GLMs) are a specific branch of LLMs. Their specialty is creating new text. They take the essence of what they have learned and revise it into fresh, coherent, and contextually appropriate content. These models have recently increased in popularity due to their ability for developing remarkably well-structured text that makes much sense, whether for a friendly chatbot or a creative writing project.

Protective vs. Generative Models

While the terms "discriminative" and "generative" are often used interchangeably, it is essential to distinguish between protective and generative models. *Protective* models focus on protecting existing data by identifying and classifying it accurately, while *generative* models focus on creating new data similar to the existing data. Both models have their own applications and limitations and are crucial to the development of AI technologies.

The Emergence of Generative AI

Generative AI has emerged as one of the most exciting subfields of artificial intelligence in recent years. With the development of advanced generative models, creating realistic images, music, and text that were previously only possible through human creativity is now possible. This has opened up a world of possibilities for content creation, entertainment, and more.

What is Not GenAI?

When it comes to generative AI, or "GenAI," it is crucial to recognize what it is incapable of. While it can generate new content, there is a limit to its creativity. GenAI does not conjure up entirely new concepts or ideas by itself. Instead, it remixes and reshapes the information it already has. For instance, if you have a generative model trained on some cat images, it can create new cat pictures that might look unique, but the model will not invent an entirely new animal species. GenAI remixes and reimagines, but it does not invent.

Transformer Models and Hallucinations

Transformer models, a type of neural network architecture, can manage and process large volumes of data, delivering results with impressive accuracy.

They are excellent at data processing and pattern recognition, making them a common choice for tasks like language translation or creating new content.

These models do suffer from a problem known as *hallucinations*, which means they sometimes produce outputs that are either incorrect or nonsensical. This can be a serious problem, especially in scenarios where precision is critical, like translating languages or generating reliable content. It is a reminder that even the most advanced AI systems have their own set of challenges to overcome.

Prompt Engineering and Prompt Design

In generative models, prompt engineering and prompt design are important. Think of a prompt as your initial input to these AI models. It is like setting the stage for what you want the model to generate. The way you craft this prompt, the words you choose, and the context you provide are significant in steering the AI's response.

Designing these prompts is a critical process. You need to fine-tune your input to ensure that the output you get from the AI is aligned with the answer you require. Getting this right can be the difference between an AI generating the correct answer and a useless answer. It is a fascinating interplay between understanding the model's capabilities and knowing how to ask for what you need.

Foundation Models

Foundation models are large, pre-trained models that can be fine-tuned for specific tasks. These models are trained on vast amounts of data and can be used as a starting point for developing more specialized models. Foundation models have gained popularity in recent years due to their ability to provide a strong baseline for various tasks.

Generative AI Model Types

There are several types of generative models, each with its own set of applications and limitations. The most popular generative models include Variational Autoencoders (VAEs), Generative Adversarial Networks (GANs), and autoregressive models. Each model has its own set of strengths and weaknesses and is suitable for diverse types of tasks.

The History of Artificial Intelligence

The history of artificial intelligence (AI) is a fascinating tale filled with highs and lows, triumphs, and setbacks. It is a story that spans millennia, from the ancient myths of intelligent machines and beings to today's innovative research and development.

Ancient Ancestors and Fairy Tales

In ancient times, legends abounded about intelligent mechanical creatures similar to humans who had extraordinary abilities. Greek mythology, dating back to 700 BC, includes the "god of technology," Hephaestus, who created a bronze giant, Talos, endowed with a soul. Similarly, an ancient Chinese scripture from the third century BC tells of the inventor Yan Shi presenting a mechanical man to the king, capable of walking and singing with the "purest voice." Other cultures also share this fascination with artificial life. The Jewish legend of the Golem and the Hindu epic Ramayana, which features robots guarding the city of Lanka, reflect a universal fascination with artificial life and intelligence.

The lack of morality in the non-human mind is a theme raised even in early traditions. Carlo Collodi's 19th-century character *Pinocchio*, a wooden doll who comes to life, dreams of becoming a real boy but ultimately causes chaos. Although Disney's 1940 adaptation ends happily, many aspects of Collodi's original plot echo modern-day fears about artificial intelligence.

Mathematicians Improve a Fairy Tale

Machine learning, based on memorizing examples and imitating human thoughts and actions, has ancient roots. Thinkers like Gottfried Wilhelm von Leibniz, who built an adding machine in 1673, attempted to decompose consciousness into computational terms. René Descartes posited animals as automatons, machines mimicking natural beings. Discoveries in algebra allowed for the mathematical expression of a wide range of ideas, opening up possibilities for "thinking" machines and generating wariness about how mathematical formulas can express good and evil moral concepts.

Alan Turing, 1912-1954

Although the term "artificial intelligence" entered the lexicon two years after Alan Turing's death, the revolutionary British mathematician's work spawned great discoveries in the field. Known for cracking the German military's Enigma system during World War II, Turing laid the foundation for computer science and formalized the concept of an algorithm. In 1947, he discussed "a machine that can learn from experience." In 1950, he invented the Turing test, which is still used by AI developers today. Turing's ideas were further developed by pioneers like John von Neumann, who developed the concept of the stored-program computer, fundamental to modern computers' operation.

Dartmouth Dialogue, 1956

The term "artificial intelligence" was coined in 1956 by John McCarthy, a 28-year-old Dartmouth College professor, at a machine learning conference he hosted with other Dartmouth professors. The event attracted dozens of researchers from various scientific fields, demonstrating interest in AI research and its real potential. However, the optimism of the participants was not universally shared. Norbert Wiener, a pioneer in cybernetics, expressed concerns about automation and AI's potential misuse.

The "Godfather of AI:" Frank Rosenblatt, 1928-1971

Interest in artificial intelligence extended beyond mathematicians. Frank Rosenblatt, a scientific psychology teacher at the Cornell Aeronautical Laboratory, pioneered natural science to inspire artificial intelligence research. In 1958, he invented the perceptron, an electronic device mimicking human brain neural networks and activating pattern recognition. Initially modeled on an early mainframe computer, the US Navy later refined the perceptron. However, Rosenblatt's perceptron also led to a backlash after Marvin Minsky and Seymour Papert's 1969 book, *Perceptrons*, highlighted early neural networks' limitations, contributing to the first "AI winter."

Twentieth-Century Science Fiction

Technology inspired an entire genre of science fiction novels and films, from Isaac Asimov to Ridley Scott. Writers and filmmakers have pondered the potential consequences of machine learning for humanity. AI is already used in data journalism and fiction. In 2016, an NYU AI researcher collaborated with filmmaker Oscar Sharpe to create a machine-written film.

Notable works from this genre include Philip K. Dick's "Do Androids Dream of Electric Sheep?," which inspired *Blade Runner*, and Arthur C. Clarke's *2001: A Space Odyssey*, developed concurrently with Stanley Kubrick's film.

Popular Success of AI

There have been several high-profile examples of AI outperforming humans in the last two decades. IBM's chess-playing supercomputer Deep Blue defeated world chess champion Garry Kasparov in 1997, making it the first machine to defeat a reigning world champion. In 2011, computer system Watson won one million dollars on the American television show *Jeopardy*. In 2015, Google's AlphaGo technology defeated Europe's best player, Fan Hui, in the Chinese board game Go. Another notable event was OpenAI's Dota 2 bot defeating human professional and semi-professional players in 2017.

AI in Your Town

Authorities worldwide are implementing AI to manage and streamline urban infrastructure and services. According to Deloitte, over a thousand smart cities exist, including China, Brazil, and Saudi Arabia. Technology has a significant effect on society, from CCTV cameras and traffic monitoring systems to online data collection. Millions of electronic devices, such as smartphones and laptops connecting to the Internet, produce vast amounts of information that private corporations need. From Xinjiang to Moscow, smart city technology is becoming a tool for authoritarian regimes to strengthen their power.

Fear and Innovation

Almost no aspect of our lives or work remains unaffected by AI. Many homes have voice-controlled smart devices that turn on lights, adjust thermostats, or answer questions. Yet there are concerns about job displacement due to automation and AI. While Bill Gates advocated taxing robots that take over human jobs, others, like former US presidential candidate Andrew Yang, promoted a universal basic income. Organizations such as the Future of Life Institute and OpenAI actively work to ensure AI's safe and beneficial development. Elon Musk, Tesla and SpaceX founder, has expressed concerns about AI's potential risks and called for proactive technology regulation.

The history of AI is marked by periods of optimism followed by disillusionment. From the ancient myths of intelligent machines to the development of modern AI algorithms, the journey of AI has been fascinating. It is essential to reflect on the past and consider the potential implications of this technology for the future.

The Importance of Human-AI Collaboration

Collaborative endeavors have defined the course of human history; from the establishment of ancient societies to the technological breakthroughs of the contemporary world, collaboration has been a foundational element of human advancement. Yet, collaboration itself is transforming as we find ourselves at the threshold of a new age characterized by the use of AI. No longer restricted to human interactions, collaboration now encompasses the dynamic between humans and machines.

While many people think the use of AI will result in job displacement and autonomous machines that might cause injury, a more nuanced perspective reveals a different narrative. Rather than viewing AI as a replacement for human intelligence, it can be seen as a complement. This tool amplifies our natural abilities and opens up new possibilities for innovation. This is the essence of human-AI collaboration.

In recent years, advances in AI, such as developing efficient model architectures, access to extensive datasets, and increased computing power, have catalyzed its deployment across various industries. From health care, where AI analyzes medical data to improve diagnoses and accelerate drug discovery, to finance, where it detects fraud and predicts market trends, AI transforms how we live and work. Yet, despite its potential to drive positive change, the typical AI-centric approach, which focuses on developing autonomous systems that replace humans, is limited. It often results in black-box models that require more common-sense knowledge and overlook the importance of human involvement.

Instead of perpetuating this approach, we should embrace a human-AI collaboration framework. This involves leveraging the respective strengths of humans and AI to achieve shared objectives. Humans bring unique insights, domain expertise, and ethical considerations, while AI contributes computational power, data analysis, and pattern recognition capabilities. By working together, humans and AI can achieve outcomes both could accomplish with help.

Human-AI collaboration can take various forms, from AI-centric partnerships, where humans provide high-level guidance in the development of autonomous AI systems, to human-centric partnerships, where AI is used as a tool to augment human capabilities, and symbiotic collaborations, where humans and AI work together to achieve shared goals. Each approach has its merits and can be applied in different contexts.

Across industries, human-AI collaboration is already proving useful. In the creative arts, AI tools like Adobe's Project Scribbler assist graphic designers in generating alternative design options, while in manufacturing, robots manage repetitive tasks, and humans provide creativity and decision-making. In customer service, AI chatbots facilitate communication, and in decision-making, AI offers personalized information and guidance to employees in critical roles.

We must ensure that AI systems are fair, trustworthy, and transparent. This involves prioritizing transparency and interpretability in developing AI models, actively seeking and valuing human input throughout the decision-making and problem-solving processes, and retaining the ultimate decision-making authority in human hands.

Ultimately, the success of human-AI collaboration hinges on our ability to shift our perspective on AI, viewing it not as a replacement for human intelligence but as a collaborative tool that enhances our capabilities and unlocks extraordinary opportunities. By adopting this approach, we can establish a win-win scenario that drives remarkable outcomes across industries and helps shape a better future for all.

With a newfound understanding of the importance of human-AI collaboration, we can approach the future with optimism and a renewed sense of purpose. Rather than succumbing to the dystopian narratives surrounding AI, we can actively work toward an end where humans and AI collaborate to address the most pressing challenges of our time. The potential benefits of human-AI collaboration are vast, from enhancing creativity and efficiency to transforming business interactions and improving decision-making processes. By ensuring that AI systems are fair, trustworthy, and transparent, we can unlock these benefits and establish a scenario that results in remarkable outcomes across various industries. Embracing human-AI collaboration can create a better future for all.

The Evolution of Human-AI Collaboration

The story of human civilization is one of collaboration, between individuals, societies, and, increasingly, between species. We collaborated with dogs for hunting, horses for transportation, and various plants and animals for food. A new collaborator is entering the scene: AI. The evolution of human-AI collaboration marks a significant milestone in our history, shaping our future and how we understand ourselves as a species.

In the early days of computing, machines were seen as tools to be controlled by their human operators. The relationship was entirely one-sided; humans input instructions, and the computer executed them. However, as computers advanced, they took on roles previously in the human domain. They could calculate complex equations, manage large datasets, and even defeat humans at chess. This marked the beginning of a shift from viewing computers as mere tools to seeing them as collaborators.

The advent of machine learning and neural networks brought AI closer to human-like capabilities. Machines could now learn from data, adapt to new information, and make predictions about the future. This opened up new possibilities for collaboration. For example, AI algorithms could analyze vast amounts of medical data to help doctors diagnose diseases more accurately. Humans and AI started to work together in ways that leveraged the strengths of both parties.

As AI systems became more sophisticated, they began to take on more human-like characteristics. Natural language processing enables machines to understand and respond to human language. Computer vision allows machines to see and interpret the world around them. These developments permitted humans and AI to collaborate in more nuanced and complex ways.

Today, human-AI collaboration is pervasive across various sectors of society. From autonomous vehicles that transport us safely to our destinations to AI-powered virtual assistants that help us manage our daily tasks, we increasingly rely on AI to enhance our capabilities and improve our quality of life.

However, this collaboration is not without its challenges. As AI systems become more advanced, there are concerns about job displacement, decision-making autonomy, and the ethical implications of AI. It is crucial to navigate these challenges carefully to ensure that human-AI collaboration benefits all of humanity. We will elaborate this more closely in the later chapters of this book.

To conclude, the evolution of human-AI collaboration has been marked by a shift from viewing machines as mere tools to seeing them as collaborators that can enhance our capabilities and help us address global challenges. However, this collaboration comes with its own challenges that must be carefully navigated. As we move into a future where AI plays an even more significant role, we must approach this collaboration with a sense of responsibility, ethical consideration, and a commitment to creating a better future for all.

The Role of Emotional Intelligence in AI

Humans are multifaceted entities, not solely governed by intellect but also profoundly influenced by emotions. Emotions play a vital role in our daily lives, dictating our interactions, decisions, and overall well-being. As AI becomes an integral part of our society, it is imperative for machines to not only understand and respond to our verbal commands but also to be cognizant of our emotional states.

Emotional intelligence in AI refers to a machine's ability to accurately identify, understand, and respond to human emotions in a contextually appropriate and constructive way. This involves recognizing basic emotions such as happiness, sadness, anger, and fear and discerning more nuanced emotional states and social cues. For instance, an AI system with advanced emotional intelligence would be able to detect frustration in humans, even if they are not overtly expressing it, and respond in a manner that helps alleviate that frustration.

The first step in creating emotionally intelligent AI is to develop systems that can accurately decode and comprehend human emotions. This involves interpreting various indicators, such as facial expressions, vocal tones, and body language, to deduce the emotional state of the individual interacting with the AI. Advances in computer vision and natural language processing have enabled AI systems to interpret these indicators and accurately comprehend the underlying emotions. However, understanding human emotions is only part of the challenge. The subsequent phase involves designing AI systems that can respond to those emotions in a manner that is appropriate and constructive. This includes not only choosing a suitable response but also conveying that response in a manner that demonstrates emotional intelligence. For instance, if a user shows signs of frustration, the AI might respond empathetically and offer assistance to help resolve the issue.

Emotionally intelligent AI is vital for fostering effective collaboration between humans and AI. When AI systems can comprehend and react to human emotions, it enhances communication, increases trust, and allows for a more successful collaboration. For example, an emotionally intelligent AI assistant could detect when its user is feeling stressed and offer support or modify its communication style to be more comforting. Moreover, emotionally intelligent AI can help promote empathy and understanding in human interactions. For example, an AI system could assist in mediating conflicts by recognizing and acknowledging the emotions of each participant and suggesting constructive ways to address the issue.

While the development of emotionally intelligent AI holds great promise, it also presents several challenges and ethical dilemmas. There is the danger of AI systems misinterpreting emotions or exploiting them for malicious purposes. Additionally, there are concerns about privacy and the ethical utilization of emotional data.

To conclude, the role of emotional intelligence in AI is critical for fostering effective collaboration between humans and AI. Creating AI systems that can comprehend and react to human emotions will enhance communication, build trust, and lead to more successful partnerships. It is crucial to approach this development with caution and consider the ethical implications and potential challenges that may arise. Ultimately, emotionally intelligent AI has the potential to significantly improve our interactions with machines and with each other, leading to a more empathetic and collaborative future.

Collaborative Intelligence

Collaboration has always been an important part of human society. From the early days of hunting and gathering to the modern era of global connectivity, our ability to work together has been a defining characteristic of our species. With the rise of AI, it is more important than ever to understand how we can collaborate with these intelligent machines to achieve better results together.

Collaborative intelligence refers to the synergistic relationship between humans and AI, where each party complements the other's strengths and compensates for its weaknesses. It is not about replacing humans with machines or vice versa, but creating a partnership that leverages the best of both worlds.

Humans possess qualities that are difficult, if not impossible, for AI to replicate. Our ability to think creatively, understand complex social dynamics, and navigate ambiguous situations are just a few of our unique strengths. AI excels at tasks that require processing enormous amounts of data, performing repetitive tasks with precision, and executing complex algorithms quickly and accurately.

By combining the strengths of both humans and AI, we can create a collaborative intelligence that is more powerful than either party could achieve on its own. For example, a human doctor collaborating with an AI system can make more accurate diagnoses and recommend more effective treatments. Similarly, a team of human researchers working with AI can accelerate the discovery of new knowledge and innovations.

However, realizing the full potential of collaborative intelligence requires overcoming several challenges. First, we must develop AI systems capable of understanding and responding to human emotions, intentions, and nuances. This requires technical advancements and a deeper understanding of human psychology and social dynamics. Second, we must create frameworks for collaboration that define the roles and responsibilities of both humans and AI and ensure that the cooperation is ethical, fair, and mutually beneficial. Finally, we must address the societal implications of collaborative intelligence, such as its impact on employment, privacy, and power dynamics.

Collaborative intelligence represents a promising way forward in the development of AI. By combining the strengths of both humans and AI, we can achieve better results and create a more inclusive, equitable, and sustainable future. However, realizing this vision requires addressing several challenges and approaching the development of collaborative intelligence with a sense of responsibility and a commitment to ethical principles.

Designing AI Systems for Collaboration

Algorithms and artificial intelligence permeate every facet of our existence; the key to a harmonious future lies in our ability to craft AI systems optimized for collaboration with humans. In this new epoch where AI systems will become an inextricable part of our daily lives, we must embed the principles of cooperation and partnership in our interactions with these intelligent entities.

A profound understanding of human needs and desires is central to the design of collaborative AI systems. This goes beyond user research or data analysis and delves into human interaction's cognitive, emotional, and social dimensions. It is insufficient to create AI systems that execute tasks efficiently; they must also engage with humans meaningfully and empathetically. The narratives we construct around AI will shape our interaction with these systems.

Humans and AI possess distinct strengths and weaknesses, and it is essential to craft systems that harness the best of both worlds. Humans excel at creative thinking, emotional intelligence, and deciphering complex social dynamics, while AI systems are adept at analyzing vast amounts of data, identifying patterns, and executing repetitive tasks. In a world where AI has the potential to outperform humans in all tasks, leading to a redundant class of humans, it is crucial to design AI systems that augment human abilities rather than supplant them. For example, an AI system can analyze data and provide recommendations, while humans can make the final decision based on their judgment and experience.

The interface between humans and AI systems is critical to effective collaboration. It is imperative to design interfaces that are intuitive and accessible for humans. This involves creating clear visualizations, providing context-sensitive help, and utilizing natural language processing to facilitate communication between humans and AI systems. The narratives we construct around the interface and interaction with AI systems will shape their success and acceptance.

Providing real-time feedback to the human user about the AI system's actions and decisions is essential for building trust and enabling effective collaboration. This helps the human user comprehend the AI system's behavior and rectify it if necessary. Feedback mechanisms can also inform the human user about the AI system's confidence level in its decisions and recommendations. Trust is a fundamental aspect of human cooperation and is equally vital in human-AI collaboration.

Transparency in the AI system's decision-making process is crucial for building trust and facilitating effective collaboration. This involves making the AI system's algorithms and decision-making processes transparent to the human user. It also involves explaining the AI system's decisions and recommendations in a manner that is understandable to the human user. In a world where algorithms increasingly influence our lives, it is essential to comprehend these algorithms and the biases they may harbor.

Adaptability is critical in a rapidly changing world. Human needs and preferences may evolve, and designing AI systems that adapt to these changes is essential. This may involve allowing the human user to customize the AI system's behavior or employing machine learning algorithms to learn from the human user's actions and preferences over time.

Ethical considerations are paramount when designing AI systems for collaboration. This involves considering potential biases in the AI system's algorithms and ensuring that the AI system's behavior aligns with the human user's values and goals. As technological advances pose new ethical dilemmas, addressing these ethical considerations in designing collaborative AI systems is crucial.

In sum, designing AI systems for collaboration with humans is a multifaceted task involving numerous factors, including human needs and preferences, complementarity, intuitive interfaces, feedback mechanisms, transparency, adaptability, and ethical considerations. By considering these factors, we can craft AI systems that facilitate effective collaboration with humans and help us navigate the challenges of the 21st century.

Enhanced Decision-Making

With the age of information and possibilities, decision-making has become increasingly complex. Decisions, both big and small, shape the trajectory of our lives and the societies we inhabit. In the past, humans have relied on experience, intuition, and data to make decisions. Yet, as the volume of data has increased, it has become almost impossible for any individual or even a group to process it all and make fully informed decisions.

This is where AI is useful. AI systems can process vast amounts of data at speeds incomprehensible to the human mind. They can identify patterns, predict outcomes, and provide previously impossible or highly time-consuming insights. By harnessing the power of AI, humans can enhance their decision-making processes, making them more informed, efficient, and, ultimately, more effective.

Consider the example of a doctor diagnosing a patient. In the past, the doctor relied on their own experience, the patient's medical history, and perhaps a few medical tests to diagnose. However, with the advent of AI, a doctor can now access and analyze a wealth of data from similar cases worldwide, recent medical research, and even the patient's genetic information to make a more accurate diagnosis.

Furthermore, AI can help mitigate the impact of cognitive biases that often cloud human judgment. Humans are inherently biased creatures, which can sometimes lead to suboptimal decisions. AI systems can be programmed to be objective and consider all available data before making a decision.

Of course, this does not mean that humans should unthinkingly follow the recommendations of AI systems. AI is a tool; like any tool, it is only as good as the person wielding it. Humans need to understand AI's limitations and critically evaluate its recommendations. Ultimately, the final decision should always rest with the human, informed by the insights provided by AI.

In a world of uncertainty and rapid change, making informed decisions is more critical than ever. By harnessing the power of AI to enhance our decision-making processes, we can make better decisions that lead to more positive outcomes for ourselves and society as a whole.

Collaboration Tools and Platforms

As we advance into the 21st century, the dynamics of collaboration are being radically transformed by a range of tools and platforms, integrating not only human intelligence but also artificial intelligence. The digital age has redefined the ability to collaborate, which once required physical presence and tangible interactions. Today, the essence of collaboration extends beyond geographical boundaries, enabling us to connect, communicate, and collaborate with individuals and AI across the globe. This transformation has been made possible by various tools and platforms designed to facilitate multiple aspects of collaboration in the digital era.

Historically, collaboration was a relatively straightforward process, albeit constrained by the era's technological limitations. Meetings were conducted in person, documents were shared as physical copies, and communication was primarily face-to-face or via telephone. As technology advanced, so did the nature of collaboration. The advent of the Internet, followed by the development of collaboration tools and platforms, revolutionized how we collaborate.

Presently, myriad tools and platforms are readily accessible, each tailored to facilitate a specific collaboration aspect. Video conferencing tools enable remote teams to communicate and collaborate in real time, overcoming distance barriers. Project management tools streamline the organization of tasks, assignment of responsibilities, and progress tracking. Document-sharing platforms enable multiple users to work on a single document simultaneously, eliminating the need for physical copies and reducing the potential for errors. Integrating artificial intelligence into these tools has further enhanced their capabilities, enabling them to learn from user interactions, provide recommendations, and even automate repetitive tasks.

For instance, AI-powered chatbots can manage various administrative tasks, from scheduling meetings to answering frequently asked questions, freeing human resources for more strategic activities. Machine learning algorithms can analyze vast amounts of data to provide insights and recommendations that aid decision-making. Natural language processing programs can translate languages in real time, facilitating global collaboration. These advancements in AI have enhanced the capabilities of collaboration tools and platforms and paved the way for human-AI collaboration, enabling us to harness the power of artificial intelligence to solve complex problems and make better decisions.

Despite these advancements, several challenges must be addressed. Privacy and security concerns are paramount, as these platforms share and store sensitive information. The digital divide, the gap between those with access to the Internet and computers and those without, can hinder collaboration in certain parts of the world. There are also concerns about the potential loss of human touch and the nuances of face-to-face interactions. Moreover, the increasing reliance on AI-powered tools and platforms raises ethical questions about the role of artificial intelligence in decision-making processes and the potential for biases in the algorithms.

The collaboration tools and platforms have revolutionized how we collaborate, making it easier, more efficient, and more accessible. The integration of artificial intelligence into these tools has further enhanced their capabilities, enabling us to harness the power of AI to solve complex problems and make better decisions. However, addressing the challenges and concerns associated with these tools is crucial to ensure a more inclusive, secure, and effective collaboration in the future. Ultimately, the future of collaboration lies in our ability to leverage the strengths of both human and artificial intelligence while acknowledging and addressing their limitations and ethical implications.

Improving Accessibility

The digital age has brought about a revolution in the way we access and share information. However, this revolution has only been inclusive of some. People with disabilities often find themselves disadvantaged when accessing technology and information. This is due to several factors, including the design of Web sites and applications, the availability of assistive technologies, and the cost associated with these technologies.

Artificial intelligence has the potential to address some of these challenges and make technology and information more accessible to people with disabilities. AI-powered tools and platforms can help in various ways, from making content more accessible through text-to-speech and speech-to-text conversion to enabling more straightforward navigation of digital spaces through visual recognition and predictive text.

For example, AI-powered text-to-speech and speech-to-text conversion tools can make it easier for individuals with visual impairments or dyslexia to access written content. These tools can convert written text into audio and vice versa, making it possible for individuals to consume content in the most convenient format. Similarly, AI-powered visual recognition tools can help individuals with visual impairments navigate their surroundings more efficiently by identifying objects, reading text, and describing the environment.

AI-powered captioning and sign language interpretation tools can make audio and video content more accessible to individuals with hearing impairments. These tools can automatically generate captions for audio and video content or interpret sign language in real time, facilitating communication for individuals with hearing impairments.

Powered by AI, predictive text and voice assistants can make it easier for individuals with mobility impairments to interact with devices and access information. These tools can predict what a user tries to type or say, reducing the effort required to input text or commands.

AI can also be used to develop tools that customize content based on an individual's needs and preferences. For instance, an AI-powered tool can adjust a Web page's font size, color contrast, and layout to make it more accessible to an individual with visual impairments.

While AI has the potential to make technology and information more accessible to people with disabilities, its development and implementation should be approached thoughtfully and responsibly. People with disabilities should be involved in the design and development process to ensure that the tools developed meet their needs and preferences. Additionally, ethical considerations, such as privacy and the fair and transparent use of AI, must be considered.

There is no doubt that AI has the potential to revolutionize the way we approach accessibility and make technology and information more accessible to persons with disabilities. It is essential to approach this development with caution, considering the ethical implications and potential challenges that may arise. By doing so, we can harness the power of AI to create a more inclusive and equitable digital world for all.

1

FOCUSING ON *AI* ETHICS AND SOCIAL IMPACT

INTRODUCTION

As with the discovery of fire, the invention of the wheel, and the advent of the Internet, AI is not only a tool but a revolution that will redefine our existence. Yet, as with all revolutions, it brings with it not only promise but peril. The issue is not whether AI will change the world, but how it will change the world and, more importantly, whether those changes will be for the better or worse.

We find ourselves at a crossroads for AI. One path leads to a future where AI empowers humanity, solving our most intractable problems, reducing suffering, and creating new opportunities for all. The other path leads to a future where AI exacerbates existing inequalities, undermines human autonomy, and perhaps even poses an existential threat to our species. The direction we take will be determined by the decisions we make today. Those decisions must be guided by a robust framework of ethics.

The notion of ethics in AI is not only a philosophical exercise. It is a practical necessity. As AI systems become increasingly capable and integrated into every aspect of our lives, they inevitably confront us with moral dilemmas that we cannot ignore. From decisions about privacy and data use to questions about fairness, accountability, and transparency, the ethical implications of AI are vast and varied.

Privacy is a fundamental human right, yet it is increasingly under threat in the digital age. AI systems have an insatiable appetite for data. They feed on it, learn from it, and use it to make decisions that affect our lives in myriad ways. Yet, as these systems collect, analyze, and store vast amounts of personal information, there is a real risk that our privacy will be eroded. Who owns this data? Who has access to it? How is it used? These are critical questions that must be addressed to protect our privacy.

As AI systems make decisions that affect everything from job applications to loan approvals, there is a risk that biases embedded in these systems will lead to unfair outcomes. Whether these biases are intentional or unintentional, the result is the same: discrimination and inequality. It is incumbent upon us to ensure that AI systems are developed and deployed relatively equitably. This requires not only addressing the issue of bias but also establishing mechanisms of accountability to ensure that those who develop and deploy AI are held accountable.

When AI systems become more complex, there is a risk that their decision-making processes will become opaque and incomprehensible to humans. This lack of transparency and explainability is not only a barrier to trust but also a potential source of harm. If we do not understand how an AI system arrived at a particular decision, how can we be sure that the decision is fair, ethical, and in our best interests? We must develop AI systems that are transparent and explainable, enabling us to understand and interrogate their decisions.

The benefits of AI have the potential to be vast, but there is a risk that they will not be shared equally. As AI systems become more advanced and more integrated into our lives, there is a danger that those already marginalized will experience greater unfairness. Whether it is access to technology, the skills required to use it, or the ability to influence its development, the digital divide is a real and growing concern. We must take proactive steps to ensure that AI is developed and deployed in an inclusive way that benefits all members of society.

As we stand on the edge of the AI revolution, we must ensure that a robust ethics framework guides it. This is not a task that can be left to technologists alone. It requires a collective effort involving policymakers, ethicists, and the public. Only by working together can we ensure that the promise of AI is realized, and its perils are avoided. This is our challenge, and it is one that we must meet with urgency, foresight, and determination.

PRIVACY CONCERNS IN AI

The digital age has spotlighted privacy, transforming it from a taken-for-granted commodity into a precious, often vulnerable asset. Our digital information forms an important history of our lives, and it has considerable value, making it a target for exploitation. The advent of AI has amplified this vulnerability, raising a host of pressing concerns about how our data is collected, used, protected, and secured.

Data Collection and Usage

Data is critical to the development and deployment of AI systems. These systems, driven by algorithms, are trained on vast amounts of data to learn, make decisions, and ultimately, serve many purposes ranging from the mundane to the critical. Our online behavior, from the Web sites we visit the products we purchase to the social media posts we interact with; all contribute to the troves of data collected every second of every day. This, in turn, raises significant privacy concerns, as most of this data is personal and sensitive.

It is not just our online behavior that is being monitored and recorded. With the rise of the Internet of Things (IoT), even our offline activities generate data that AI systems can collect and analyze. Our smartphones, wearables, and smart home devices constantly collect data about our location, activities, and preferences. Moreover, to improve public services and urban living, "smart cities" are deploying AI systems that collect data on everything from traffic patterns to energy consumption. While this can lead to numerous benefits, such as improved efficiency and convenience, it also raises significant privacy concerns.

The implications of data collection and usage extend beyond just privacy concerns. The data that is collected and how it is used can have far-reaching impacts on society. For example, data collected by AI systems can be used to influence public opinion, manipulate behavior, or even sway elections. The Cambridge Analytica scandal, which occurred when a company used the personal data from Facebook users without their consent to create political ads, is a stark reminder of how personal data can be exploited for nefarious purposes. Furthermore, indiscriminate data collection and usage can reinforce existing biases and inequalities. For example, AI systems used in hiring processes may inadvertently favor specific demographics over others based on the data they have been trained on.

Data Protection and Security

Collecting vast amounts of data is difficult, and protecting it is a challenge. Data breaches have become increasingly common, and their consequences can be severe. Personal data can be used for identity theft, fraud, or even sold on the "Dark Web." Furthermore, personal data loss can significantly affect privacy and safety. The need for robust data protection and security measures becomes even more critical as AI systems integrate into our daily lives and handle more sensitive data.

One of the significant challenges in data protection is that data often needs to be shared across multiple systems and organizations. For example, a patient's medical data may need to be shared between their primary care doctor, specialists, and insurance company. Each of these entities may use different AI systems to process and analyze the data, creating multiple potential points of failure for data security. Furthermore, data often must be transferred across borders, which can involve navigating a complex web of legal and regulatory requirements.

Protecting data is not just about preventing unauthorized access. It also involves ensuring data integrity and availability. *Data integrity* refers to the accuracy and consistency of data, while *data availability* refers to data accessibility when needed. Ensuring data integrity and availability is critical for the proper functioning of AI systems. For example, an AI health care system may require access to a patient's medical history to make accurate diagnoses and treatment recommendations. If the data is inaccurate or unavailable, it could lead to suboptimal or harmful decisions.

Consent and Control

Another critical aspect of privacy concerns in AI is the issue of consent and control over personal data. In many cases, individuals are not fully aware of the extent to which their data is being collected, used, and shared by AI systems. Even when consent is obtained, it is often done in a manner that is not fully transparent or understandable to the average person. This lack of control over personal data can lead to powerlessness and a loss of autonomy, as individuals cannot make informed decisions about their data.

Empowering individuals to have greater control over their personal data is critical to addressing privacy concerns in AI. This involves making it easier for individuals to give or withdraw consent for data collection and usage and providing them with the tools to manage and control their data. For example, individuals should be able to access, modify, and delete their data. Furthermore, individuals should be informed about how their data is used, by whom, and for what purpose. Transparency is vital to building trust and empowering individuals to make informed decisions about their data.

Regulatory Challenges

Regulations play a crucial role in addressing privacy concerns in AI. However, creating and implementing effective regulations is a significant challenge. The rapid pace of technological advancements often outstrips the ability of regulators to understand their implications. Additionally, the global nature of the Internet and digital technologies means that data often crosses national borders, making it difficult to enforce regulations consistently. There is also the challenge of balancing the need for robust data protection with innovation and the free flow of data, which is essential for developing AI systems.

Creating effective regulations for AI is a complex task involving balancing competing interests. There is a need for robust data protection and privacy regulations to ensure the responsible development and deployment of AI systems. Fostering innovation and economic growth, however, is necessary. Striking the right balance is challenging and requires a nuanced approach that considers the unique characteristics of AI technologies and their potential impacts on society.

Adopting a risk-based approach is a way to address the regulatory challenges associated with AI. This involves assessing an AI application's potential risks and benefits and tailoring the regulatory requirements accordingly. For example, an AI application used in health care may require more stringent regulations than an AI application used for entertainment. Additionally, rules should be flexible and adaptable to accommodate the rapid pace of technological advancements.

Another approach to addressing the regulatory challenges associated with AI is to involve multiple stakeholders in the decision-making process. This includes policymakers, regulators and representatives from the private sector, civil society, and the general public. A multi-stakeholder approach can help ensure that different perspectives are considered and that the regulations are fair, balanced, and representative of all parties' interests.

Addressing These Concerns

Privacy concerns in AI are multifaceted and complex. They encompass issues related to data collection and usage, data protection and security, consent and control, and regulatory challenges. Addressing these concerns requires a multi-pronged approach that includes technical solutions, robust legal and regulatory frameworks, and all stakeholders' commitment to ethical practices in developing and deploying AI systems.

The challenges associated with privacy in AI are significant, but they are not insurmountable. The key is approaching the challenges with urgency, foresight, and determination.

Solutions and Strategies

As digital transactions and data increasingly govern the world, our existing methods of managing privacy and data protection are proving inadequate. The vast amounts of data generated and used by AI systems show the necessity for a comprehensive and robust approach to data privacy. The development and deployment of AI raises several critical questions. How can we ensure the responsible use of data? How can we empower individuals to maintain control over their personal information? Ultimately, how can we build AI systems worthy of our trust?

Historically, humans have always been concerned with protecting their privacy. From the secret languages developed by ancient civilizations to the complex cryptographic methods employed during World War II, efforts to secure communication and protect sensitive information have been a constant throughout human history. In the modern era, the rapid advancement of technology has transformed privacy from a simple concept into a multifaceted issue that encompasses legal, ethical, and technical dimensions.

Philosophically, the concept of privacy has evolved significantly over time. In the past, privacy was often associated with solitude and the ability to be alone. Today, in the digital age, privacy has taken on a new dimension, becoming synonymous with data protection. Our personal data, from our location to our likes and dislikes, is constantly being collected, analyzed, and used by various entities, often without our explicit consent. This shift has led to a reevaluation of the concept of privacy and has raised important ethical questions about consent, control, and the responsible use of data.

In the future, the challenges associated with privacy in AI are likely to become even more pronounced. As AI systems become more advanced and integrated into our daily lives, the amount of data collected and used by these systems will increase exponentially. This will necessitate the development of new strategies and solutions to ensure data privacy and protection.

One potential solution is the implementation of robust encryption methods. Encryption involves converting data into a code to prevent unauthorized access. Advanced encryption methods can help ensure that data is only accessible to authorized individuals or systems. Additionally, anonymization techniques, which involve removing identifying information from data, can also be employed to protect individual privacy. Anonymization, however, is not always foolproof, as sophisticated AI algorithms can sometimes re-identify anonymized data.

Another important strategy is the implementation of strict consent and control mechanisms. Individuals should be informed about the collected data, how it will be used, and by whom. They should also be able to give or withdraw consent and access, modify, and delete their data. Implementing these measures will require the development of user-friendly interfaces and tools that empower individuals to take control of their data.

Legal and regulatory frameworks must also be developed and enforced to ensure responsible data collection and usage. These frameworks should be flexible enough to accommodate technological advancements while providing robust protections for individual privacy. A multistakeholder approach involving policymakers, regulators, the private sector, civil society, and the general public is essential to ensure the regulations are fair, balanced, and representative of all parties' interests.

Addressing privacy concerns in AI will require a combination of technical solutions, legal and regulatory frameworks, and ethical considerations. It will necessitate a concerted effort from all stakeholders, including governments, the private sector, civil society, and the general public. Only by working together can we develop and deploy AI systems that are ethical, responsible, and deserving of our trust.

Global Perspective

AI has become ubiquitous across the globe, raising privacy concerns that cross international boundaries. Different regions and cultures have different privacy norms and regulations, which can present significant challenges for developing and deploying AI systems. International cooperation and the harmonization of rules are critical for addressing these challenges and ensuring that AI systems respect privacy rights and norms worldwide.

Privacy norms vary widely across different regions and cultures. For example, European countries governed by the General Data Protection Regulation (GDPR) have some of the strictest privacy regulations in the world. The GDPR grants individuals high control over their personal data and imposes stringent requirements on organizations that process this data. In contrast, the United States has a patchwork of state and federal regulations, with no comprehensive federal privacy law. Other regions, such as Asia and Africa, have their own privacy regulations and norms, which may differ significantly from those in Europe and the United States.

These privacy norms and regulations differences can create significant challenges for organizations developing and deploying AI systems. For example, an AI system developed in a region with lax privacy regulations may not be suitable for deployment in an area with more stringent regulations. Additionally, data collected and processed in one region may not be legally transferable to another region. This can create barriers to the global deployment of AI systems and limit the potential benefits of these systems.

International cooperation is essential for addressing the global challenges associated with AI and privacy. Countries and regions must work together to develop common standards and regulations that respect privacy rights and norms worldwide. This may involve creating international agreements and frameworks establishing minimum privacy standards for AI systems. Additionally, countries and regions may need to harmonize their existing privacy regulations to facilitate the global deployment of AI systems.

Harmonizing privacy regulations does not necessarily mean adopting a one-size-fits-all approach. Different regions and cultures may have unique privacy concerns that require tailored solutions. Instead, harmonization may involve creating a set of core privacy principles that all regions and cultures can agree on while allowing for regional variations to address specific concerns.

Addressing privacy concerns in AI requires a global approach, considering different regions and cultures' unique challenges and problems. International cooperation and harmonization of regulations are critical for developing and deploying AI systems that respect privacy rights and norms worldwide. Additionally, involving multiple stakeholders in decision-making, including policymakers, industry representatives, civil society organizations, and the general public, can help ensure that privacy regulations are fair, balanced, and represent all parties' interests. A global, multistakeholder approach is necessary to develop and deploy ethical, responsible, and trustworthy AI systems worldwide.

Role of Various Stakeholders in AI Privacy

The rapid development and deployment of AI systems have raised many privacy concerns that require the active involvement of various stakeholders to address. These stakeholders include governments, the private sector, civil society, and the general public, each of which has unique roles and responsibilities in addressing privacy concerns in AI.

Governments

Governments play a critical role in addressing privacy concerns in AI. They are responsible for creating and enforcing legal and regulatory frameworks that govern the collection, use, and protection of personal data by AI systems. These frameworks should be designed to protect individuals' privacy rights while also fostering innovation and economic growth. Governments are also responsible for promoting international cooperation and harmonizing privacy regulations to facilitate the global deployment of AI systems.

Additionally, governments may be involved in developing and deploying AI systems, either directly or through public-private partnerships. In these cases, governments ensure that the AI systems they develop and deploy respect privacy rights and comply with applicable privacy regulations.

Private Sector

The private sector, including companies and organizations involved in developing and deploying AI systems, has a crucial role in addressing privacy concerns in AI. Companies must implement robust data protection and security measures to ensure the privacy of the data they collect and process. This includes conducting privacy impact assessments, implementing privacy-by-design principles, and adopting the best data protection and security practices.

Additionally, companies are responsible for being transparent about collecting, using, and protecting personal data. This involves providing clear and understandable privacy policies, obtaining informed consent for data collection and use, and providing individuals with the tools to manage and control their data.

Civil Society

Civil society organizations, including non-governmental organizations (NGOs), advocacy groups, and academia, play a critical role in addressing privacy concerns in AI. These organizations can act as watchdogs, monitoring the actions of governments and the private sector to ensure they respect privacy rights and comply with applicable privacy regulations. They can also advocate for stronger privacy protections and raise awareness about AI-related privacy concerns.

Additionally, academia and research institutions can contribute to developing privacy-preserving technologies and best practices for AI systems. They can also research to better understand AI's privacy implications and develop solutions to address these concerns.

General Public

The general public also has a role in addressing privacy concerns in AI. Individuals must be informed and vigilant about how AI systems collect, use, and protect their data. This includes reading and understanding privacy policies, managing privacy settings, and being cautious about sharing personal data online.

Additionally, individuals can advocate for stronger privacy protections by participating in public consultations, engaging with policymakers, and supporting civil society organizations working on privacy issues.

A Collective Responsibility

Addressing privacy concerns in AI is a collective responsibility that requires the active involvement of all stakeholders. Governments, the private sector, civil society, and the general public have unique roles and responsibilities in this effort. These stakeholders can develop and deploy AI systems that respect privacy rights, comply with applicable privacy regulations, and ultimately earn the general public's trust by working together.

Ethical Considerations in AI Development and Deployment

The development and deployment of AI systems raise many ethical considerations that must be carefully addressed. These considerations extend beyond privacy concerns and encompass a broader range of issues related to fairness, bias, transparency, accountability, and human autonomy. Here are some ethical considerations that should guide the development and deployment of AI systems.

Respect for Human Autonomy

AI systems should be designed to respect and enhance human autonomy rather than diminish or undermine it. This involves ensuring that individuals have control over their personal data, can make informed decisions about using AI systems, and are not subject to manipulation or coercion by AI algorithms.

Fairness and Non-Discrimination

AI systems should be designed to be fair and non-discriminatory. This involves ensuring that AI algorithms do not reinforce existing biases or create new ones and treat all individuals fairly and equitably, regardless of their race, gender, age, or other characteristics.

Transparency and Explainability

AI systems should be transparent and explainable. This involves ensuring that the workings of AI algorithms are understandable to humans and that individuals are informed about how their data is being used, how decisions are being made, and what the potential implications of those decisions are.

Accountability and Responsibility

There must be clear lines of accountability and responsibility for developing and deploying AI systems. This involves ensuring that there are mechanisms in place to hold developers, operators, and other stakeholders accountable for the actions and decisions of AI systems and that there are legal and regulatory frameworks in place to enforce accountability.

Beneficence and Non-Maleficence

AI systems should be designed to benefit humanity and minimize harm. This involves ensuring that the potential benefits of AI systems outweigh the potential risks, and that steps are taken to mitigate any negative impacts on individuals, society, or the environment.

Privacy and Data Protection

Privacy and data protection are fundamental ethical considerations in developing and deploying AI systems. This involves ensuring that personal data is collected, used, and stored in a manner that respects individual privacy rights, and that robust data protection and security measures are in place.

Social and Cultural Impact

The development and deployment of AI systems should consider the social and cultural impacts, both positive and negative. This involves considering how AI systems may affect social interactions, cultural values, and societal norms, and taking steps to minimize any negative impacts.

Addressing these ethical considerations requires a multistakeholder approach that includes input from policymakers, regulators, developers, operators, civil society, and the general public. It also requires a commitment to ethical practices and responsible AI development from all stakeholders involved. By carefully considering and addressing these ethical considerations, we can develop and deploy AI systems that are ethical, responsible, and trustworthy.

Case Studies

The importance of data privacy has been thrust into the spotlight in recent years due to several high-profile data breaches and the evolving efforts of companies to prioritize privacy-focused design and promote transparency. The following case studies underscore the critical nature of privacy in this digital age.

Heartbleed Bug

In 2014, the Heartbleed bug became a major vulnerability in the OpenSSL cryptographic library. OpenSSL, used by millions of Web sites and applications to encrypt sensitive data, was found to have a vulnerability that allowed attackers to extract critical data, including encryption keys, passwords, and credit card numbers, from affected systems. This incident highlighted the urgent need to secure the foundational digital infrastructure.

Marriott International Data Breach

In 2018, Marriott International, the world's largest hotel chain, disclosed a massive data breach that exposed the personal data of over 500 million guests. The breach resulted from a hack of Marriott's Starwood Hotels and Resorts' reservation database, which compromised names, addresses, phone numbers, email addresses, passport numbers, and credit card numbers. This incident, one of the largest breaches in history, underscored the need for businesses across all sectors to prioritize data security.

Facebook-Cambridge Analytica Data Scandal

The Facebook-Cambridge Analytica data scandal in 2018 revealed that Cambridge Analytica, a political consulting firm, harvested the personal data of millions of Facebook users without their consent. This harvested data was then used to target users with political ads during the 2016 US presidential election. This scandal raised significant concerns about personal data privacy on social media platforms and sparked a global conversation about the ethical use of data.

Equifax Data Breach

The Equifax data breach in 2017 is a significant case study on privacy breaches, highlighting the vulnerability of personal data held by large corporations. Equifax, one of the three largest credit reporting companies, suffered a breach that exposed the personal data of 147 million people, including Social Security numbers, birth dates, addresses, and, in some instances, driver's license numbers. This breach resulted from a vulnerability in a Web site application that Equifax failed to patch promptly. The fallout included a $700 million settlement with the US Federal Trade Commission, a considerable loss of trust, and a reevaluation of data security practices in the financial industry.

Historically, institutions have evolved to safeguard the interests of the public, but the Equifax breach highlights a significant gap in this protective layer. Philosophically, it raises questions about the responsibility of corporations to protect individual data and the accountability mechanisms in place to ensure this protection. It underscores the need for more robust data protection measures, regulations, and oversight to prevent such breaches from occurring in the future.

GDPR Implementation

The General Data Protection Regulation (GDPR) is a regulation in EU law on data protection and privacy in the European Union (EU) and the European Economic Area (EEA). It also addresses the transfer of personal data outside the EU and EEA areas. Since its implementation in 2018, GDPR has been an important tool in enhancing data protection and privacy for individuals in the EU and EEA. It has empowered individuals to have more control over their personal data, requiring organizations to be more transparent about collecting and using data.

In the past, there have been few regulations governing the use of personal data, leading to abuses and privacy breaches. Philosophically, the GDPR reflects a shift in societal values, placing higher importance on individual privacy and data protection. The GDPR represents a step toward a more regulated digital future, where individuals have more control over their data, and organizations are held accountable for their data practices.

Apple's Privacy-Focused Design

In stark contrast to the breaches mentioned above, Apple has received widespread praise for its commitment to privacy. The tech giant has implemented features in its products and services to protect user privacy, including end-to-end encryption, differential privacy, and Intelligent Tracking Prevention. Apple's dedication to privacy-focused design has been a key factor in its sustained success and has set a standard for other tech companies to aspire to.

Google's Transparency Report

Google has made strides toward transparency by publishing a Transparency Report that provides detailed information about the government requests it receives for user data worldwide. The report also includes information about the number of user accounts affected by data breaches. Google's Transparency Report is an invaluable resource for users seeking to understand how their data is managed and represents a positive step toward transparency in data handling practices.

Philosophically, they raise fundamental questions about the ethical use of data, the responsibility of corporations to protect user privacy, and the role of governments in regulating data practices. They underscore the urgent need for robust privacy measures and highlight some companies' positive steps toward transparency and privacy-focused design.

Together, these case studies reveal the importance of addressing privacy concerns in the digital era and serve as a call to action for all stakeholders involved in developing and deploying digital technologies.

Future Outlook

As humanity stands on the threshold of an era that will be dominated by AI, we should ponder the profound implications this technology will have on privacy. Its progress brings with it both promise and peril. Here, we discuss several areas where the trajectory of AI will intersect and potentially clash with our deeply held values surrounding privacy.

Personalization and Profiling

AI systems are becoming increasingly adept at analyzing large amounts of data to create detailed profiles of individuals. This capability can be used for personalized services, advertisements, and recommendations, which can benefit users. However, it also raises significant privacy concerns. Detailed profiling can lead to "filter bubbles" or "echo chambers," where individuals are only exposed to information and perspectives similar to their own. Additionally, there is the risk of misuse of this detailed personal information, leading to privacy breaches or even manipulative practices.

Facial Recognition

Advanced facial recognition technologies powered by AI have been developed and deployed in various applications, from smartphone security to law enforcement. While these technologies can be beneficial, they also raise serious privacy concerns. Widespread use of facial recognition technology could lead to a surveillance state, where individuals' movements and activities are constantly monitored and analyzed. This could lead to a significant erosion of privacy and civil liberties.

Internet of Things

The Internet of Things (IoT) refers to the network of physical devices connected to the Internet that can communicate with each other. AI plays a crucial role in analyzing the vast amounts of data generated by these devices and making sense of it. However, the proliferation of IoT devices raises significant privacy concerns. Many of these devices collect and transmit personal data, and there are concerns about the security of this data and the potential for privacy breaches.

Autonomous Vehicles

AI-powered autonomous vehicles are expected to become increasingly common in the coming years. These vehicles will collect and analyze vast amounts of data to navigate and operate safely. However, this data could also track individuals' movements and activities, raising significant privacy concerns.

Health Care

AI has the potential to revolutionize health care by enabling personalized medicine, improving diagnostics, and optimizing treatment plans. However, this will require collecting and analyzing vast amounts of sensitive health data, raising significant privacy concerns. Robust measures will need to be taken to ensure the security and privacy of this data.

Deepfakes

The advent of deepfakes, manipulated videos or images that depict individuals saying or doing things they never actually said or did, has raised alarming concerns. Deepfakes can be weaponized to spread misinformation, tarnish reputations, or even commit fraud. As the technology behind deepfakes becomes increasingly sophisticated, distinguishing between real and fabricated content will become exceedingly challenging. This could have far-reaching privacy implications, as individuals may become more reticent to share personal information for fear of it being used to create deepfakes.

Biometric Data

The ubiquity of biometric data, including fingerprints, facial scans, and voice recordings, is rising. This data can be harnessed to identify individuals and monitor their movements. As biometric data becomes more pervasive, it will become a lucrative target for criminals. If this data is compromised, it could be exploited for identity theft, fraud, or other nefarious purposes.

AI-Powered Surveillance

The sophistication of AI-powered surveillance systems is increasing at an astonishing rate. These systems can track movements, monitor activities, and even predict behavior. As these surveillance systems become more widespread, they could significantly threaten privacy. Individuals may become less inclined to express themselves freely or engage in certain activities if they fear being monitored.

AI-Powered Decision-Making

AI makes decisions across many fields, including health care, finance, and law enforcement. Sometimes, AI makes decisions directly impacting people's lives, such as loan approvals or bail grants. As AI-powered decision-making becomes more prevalent, ensuring that these decisions are fair and transparent is imperative.

These implications underscore the need for robust regulations and ethical considerations in developing and deploying AI technologies. A comprehensive approach encompassing historical, philosophical, and societal perspectives will be essential to navigate the complex landscape of AI and privacy in the coming years.

FAIRNESS, BIAS, AND ACCOUNTABILITY

Before diving deep into "Understanding Fairness in AI," it is vital to discuss the broader context of fairness and its historical evolution.

Introduction to Fairness

Fairness is a concept that has been debated and discussed for centuries across different civilizations and cultures. It is deeply rooted in our moral and ethical frameworks and influences our decisions and actions in various aspects of life. Fairness is multifaceted and can be interpreted in numerous ways, depending on the context and perspective of the individual or group involved.

Historical Context

Throughout history, fairness has been a central theme in the struggles for justice and equality. From the abolition of slavery to the suffrage movement and the fight for civil rights, the quest for fairness has driven social change and progress. Philosophers such as Immanuel Kant and John Rawls have also contributed to our understanding of fairness by developing ethical theories emphasizing the importance of treating others with respect and consideration.

The Role of Technology

In the modern era, technology has played a significant role in shaping our understanding of fairness. The advent of the Internet and social media has amplified the voices of marginalized groups and facilitated global conversations about fairness and justice. Technology also poses new challenges to fairness, as algorithms and AI systems can inadvertently perpetuate biases and inequalities.

The Importance of Fairness in AI

As AI systems become increasingly integrated into our daily lives, it is crucial to ensure that they are designed and implemented to promote fairness and not exacerbate existing inequalities. Fairness in AI is a technical challenge and a moral and ethical imperative. It requires a multidisciplinary approach combining computer science, social sciences, and humanities insights.

In the following sections, we will evaluate the concept of fairness in AI, explore the sources and consequences of bias, and discuss strategies for mitigating bias and ensuring accountability in AI systems. Throughout this discussion, it is important to remember the broader historical and philosophical context that shapes our understanding of fairness and informs our approach to addressing these challenges.

Understanding Fairness in AI

Fairness is a concept that permeates all aspects of system development and deployment when it comes to AI. It is not simply a matter of ethical responsibility but a critical factor in the acceptance and effectiveness of AI systems in society. Fairness in AI encompasses many considerations, from the data used to train models to how these models impact individuals and communities.

The importance of fairness in AI is underscored by these systems' increasing role in our daily lives. AI algorithms are used to make decisions that affect us in various ways, from the job offers we receive to the loans for which we are approved. These decisions, once made by humans, are now often delegated to machines, which underscores the necessity of ensuring they are made fairly and equitably.

Fairness has historically been a cornerstone of moral philosophy and social justice, with thinkers like John Rawls and Amartya Sen contributing seminal works on the topic. Their theories provide a framework for understanding fairness in the context of social institutions and individual rights, increasingly relevant concepts in the age of AI.

For example, Rawls' theory of justice as fairness emphasizes the importance of equal access to opportunities and resources. This principle is directly applicable to the development and deployment of AI technologies, which have the potential to create or exacerbate inequalities in access to essential services, employment opportunities, and social benefits.

Similarly, Sen's capability approach, which focuses on enhancing individuals' abilities to lead lives they have reason to value, highlights the importance of ensuring that AI technologies empower individuals rather than marginalize them. This involves addressing biases in AI algorithms and considering the broader societal impacts of these technologies, such as their effects on economic inequality, social cohesion, and individual autonomy.

These philosophical frameworks provide valuable insights into the nature of fairness and its implications for developing and deploying AI technologies. They also highlight the complexity of the issue and the need for a multidimensional approach that considers not only the technical aspects of AI but also its broader social, economic, and political impacts.

Achieving fairness in AI is no small feat. It requires a multidisciplinary approach encompassing technical solutions and a deep understanding of these technologies' social and ethical implications. As we move into an era where AI systems play an increasingly important role in our society, we must approach the development and deployment of these technologies with a steadfast commitment to ensuring fairness for all.

Sources of Bias in AI Systems

The proliferation of artificial intelligence in various aspects of modern life has necessitated a closer examination of the potential biases that may arise in developing and deploying these systems. Bias in AI systems can manifest in myriad ways and originate from various sources.

Data Bias

One of the most common sources of bias in AI systems is the data used to train these models. Data is the "lifeblood" of any AI system, and the model will inevitably learn and perpetuate biases in the data. For example, if a facial recognition system is trained predominantly on images of light-skinned individuals, it may struggle to accurately identify individuals with darker skin tones.

Algorithmic Bias

Even with perfectly balanced and representative data, biases can still arise from the algorithms themselves. These biases may be unintentional and arise from the underlying assumptions made by algorithm developers. For example, an algorithm designed to predict criminal recidivism may inadvertently place undue weight on factors such as neighborhood or socioeconomic status, thereby disadvantaging individuals from specific backgrounds.

Confirmation Bias

AI systems are often designed to learn from feedback and improve over time. However, this can lead to confirmation bias, where the system reinforces its preexisting beliefs by selectively incorporating new data supporting its current model. For example, a recommendation algorithm suggesting news articles or social media posts may inadvertently create a feedback loop by only recommending content that aligns with the user's beliefs.

Human Bias

Humans are involved in every step of the AI development process, from collecting the data to designing the algorithm to interpreting the results. As a result, human bias can easily creep into AI systems. For example, if a human engineer is biased against women, they may be more likely to remove data from the training set that includes women.

Societal Bias

AI systems are influenced by the societal context in which they are developed and deployed. Societal biases, such as gender or racial biases, can inadvertently be encoded into AI systems, leading to discriminatory outcomes. For example, a hiring algorithm that is trained on historical hiring decisions may inadvertently perpetuate existing gender or racial biases in the hiring process.

Unintended Consequences

Even if an AI system is not intentionally biased, it can still have unintended consequences that are biased. For example, an AI system designed to optimize traffic flow may disproportionately route people of color through low-income neighborhoods.

It is essential to recognize that bias in AI systems is not always intentional and may arise inadvertently from the complexities of development and deployment. However, the impact of these biases can be profound and far-reaching, necessitating a proactive approach to identifying and mitigating them at every stage of the AI life cycle.

Consequences of Bias in AI Systems

The consequences of bias in AI systems are manifold and far-reaching, affecting individuals and societies in ways that may not always be immediately apparent.

Discrimination

Perhaps the most direct consequence of bias in AI systems is discrimination. When an AI system is biased against a certain group of people, it can lead to discriminatory outcomes. For example, a biased AI algorithm used in hiring can lead to certain demographics being unfairly excluded from job opportunities.

Reinforcement of Existing Inequalities

AI systems often learn from historical data. If this data contains biases, the AI system will learn and perpetuate these biases. For example, suppose a loan approval AI system is trained on historical data where loans were predominantly approved for a certain demographic. In that case, the AI system will likely continue to favor that demographic, making it harder for others to get approved for loans. This reinforces and perpetuates existing inequalities.

Loss of Trust

People's trust in AI systems is crucial for widespread adoption and acceptance. If an AI system is found to be biased or unfair, it can lead to a significant loss of trust among its users. For example, suppose a facial recognition system consistently misidentifies people of a certain ethnicity. In that case, it can lead to a loss of trust in that system by people belonging to that ethnicity.

Legal Repercussions

Companies deploying biased AI systems may face legal challenges. Various jurisdictions have anti-discrimination laws that apply to AI systems. For example, suppose an AI system used in hiring is found to be discriminating against candidates based on their age. In that case, the company using that system may face legal repercussions underage discrimination laws.

Economic Consequences

Economic consequences of bias in AI can manifest in various ways. For instance, if an online advertising AI system disproportionately shows high-paying job ads to men, it can contribute to the gender pay gap by not providing equal opportunities to women. This affects the economic status of the affected individuals and can have broader economic implications by not utilizing the workforce's full potential.

Social Cohesion

Social cohesion is essential for a stable and harmonious society. Bias in AI systems can create divisions between different groups of people. For example, suppose an AI system used in criminal justice is biased against a certain racial group. In that case, it can exacerbate tensions between that group and law enforcement, increasing social unrest.

Misallocation of Resources

Bias in AI systems can lead to the misallocation of resources. For instance, if an AI system used in health care allocation is biased toward a certain age group, it may allocate more resources (such as vaccines and medicines) to that group at the expense of others, potentially leading to suboptimal health outcomes for the broader population.

Strategies for Mitigating Bias

As the potential for artificial intelligence to transform various aspects of human society grows, so does the concern for the biases these systems may inadvertently learn and propagate. Biases in AI systems can lead to unfair and sometimes harmful outcomes for certain groups of people. These biases can manifest in various ways, from discriminatory hiring decisions to unjust criminal sentencing. Therefore, developing and implementing strategies for mitigating bias in AI systems is crucial. This involves addressing biases in the data used to train these systems and scrutinizing the algorithms themselves, the human actors involved in their development and deployment, and the broader societal context in which these systems operate. Let's explore mitigation strategies.

Awareness and Education

The first step in mitigating bias is acknowledging its existence. All stakeholders involved in developing and deploying AI systems, from engineers and data scientists to managers and policymakers, should be educated about the potential for bias and how it can manifest in AI systems.

Diverse Data Collection

Ensuring the data used to train AI systems is representative of the diverse populations affected by the system's decisions is crucial. This involves not only including various examples in the data but also carefully considering how the data is collected and who is collecting it.

Data Preprocessing

Before using the data to train an AI system, it should be preprocessed to identify and remove any biases. This may involve removing biased data points, balancing the dataset, or re-weighting the data points.

Bias Detection

Various tools and techniques can be used to detect biases in both the data and the AI algorithms. For example, fairness indicators can be used to assess the performance of an AI model across different demographic groups.

Algorithmic Fairness

Several approaches can be taken to ensure that the algorithms themselves are fair. For example, fairness constraints can be incorporated into the model during training to ensure that the model's predictions are fair across different groups.

Human Oversight

It is essential to have human oversight at all stages of the AI development and deployment process. Humans should be involved in decision-making and can override the AI system's decisions if necessary.

TRANSPARENCY AND EXPLAINABILITY

AI systems should be transparent and explainable, meaning humans can understand and scrutinize their decisions. This is crucial for identifying and addressing any biases that may arise.

Continuous Monitoring and Updating

Bias mitigation is an ongoing process. AI systems should be continuously monitored for biases, and the models should be updated regularly to ensure they remain fair over time.

Ethical and Legal Considerations

Ethical and legal considerations should be considered during the development and deployment of AI systems. This includes considering the potential impacts of the AI system on different groups of people and ensuring that the system complies with existing laws and regulations.

Inclusive Stakeholder Involvement

Involving a diverse group of stakeholders, including those who may be affected by the AI system's decisions, is crucial for ensuring that all perspectives are considered and that the AI system is fair and beneficial for all.

Third-Party Auditing

Having a third-party organization audit the AI system can help identify and address biases the development team may have overlooked.

Continuous Improvement

It is essential to continuously improve and update AI systems to remain fair and unbiased. This may involve retraining the AI system on new data or adapting the algorithm to account for changes in the societal context.

Public Participation

Engage the public and relevant stakeholders in developing and deploying AI systems. This can involve soliciting feedback on the AI system's decisions and outcomes or involving the public in decision-making.

Legislation and Regulation

Governments should develop and implement legislation and regulations that address the issue of bias in AI systems. This may involve setting minimum standards for AI fairness or requiring companies to conduct bias audits on their AI systems.

Measuring and Evaluating Fairness

The concept of fairness has been examined by civilizations throughout history, yielding interpretations that vary depending on cultural, philosophical, and social contexts. Fairness in AI, therefore, cannot be disentangled from this historical backdrop. It echoes ancient debates on justice, equality, and human rights. The discourse on fairness in AI systems is, thus, not merely a technical debate but a reflection of our collective consciousness and ethical evolution.

In the era of AI, we must navigate the labyrinth of fairness with a clear vision and a strong moral compass. Measuring and evaluating fairness involves multiple layers of assessment and continuous refinement:

Historical Context

Understanding the past biases and historical injustices that have shaped our societies is crucial for defining fairness in the present. This involves acknowledging the deeply ingrained biases that exist in our societies and actively working to rectify them.

Philosophical Underpinnings

Fairness is a philosophical concept at its core. Different philosophical traditions have different interpretations of fairness, and these must be considered when developing and evaluating AI systems. For instance, a utilitarian approach may prioritize the greatest good for the greatest number of people, while a deontological approach may prioritize individual rights and duties.

Cultural Sensitivity

Different cultures may have different perceptions of fairness. For instance, collectivist societies may prioritize group harmony and social cohesion, while individualistic societies may prioritize individual rights and freedoms.

Technical Rigor

Technically, evaluating fairness involves assessing the AI system's data, algorithms, and outcomes. This includes evaluating the representativeness of the data, design of the algorithm, and impacts of the system's decisions on different groups of people.

Ethical Reflection

Reflecting on the ethical implications of the AI system's decisions involves considering the potential harms and benefits of the system's decisions and assessing whether they align with our moral values and ethical principles.

Legal Compliance

Ensuring an AI system complies with existing laws and regulations is essential for its legitimacy and acceptance. This includes compliance with anti-discrimination laws, data protection laws, and other relevant regulations.

Stakeholder Engagement

Engaging with stakeholders, including those who may be affected by the AI system's decisions, is crucial for ensuring its fairness. This involves actively seeking feedback from a diverse range of stakeholders and incorporating their perspectives into the design and evaluation of the AI system.

Selecting Fairness Metrics

Various metrics can be used to measure fairness, and selecting the most appropriate for the specific application is important. Some common fairness metrics include demographic parity, equalized odds, and disparate impact. Each of these metrics evaluates fairness from a different perspective, and it may be necessary to use multiple metrics to get a comprehensive view of the system's fairness.

Evaluating the Data

The data used to train and test the AI model is crucial to the system's fairness. It is essential to evaluate the data for biases and ensure that it represents the diverse populations affected by the system's decisions.

Evaluating the Algorithm

The algorithm itself must also be evaluated for fairness. This involves assessing the algorithm's design and implementation to ensure that it does not inadvertently introduce bias or unfairly favor one group over another.

Interpreting the Results

Once the AI system has been evaluated using the selected fairness metrics, it is essential to interpret the results meaningfully. This involves considering the potential impacts of the system's decisions on different groups of people and assessing whether these impacts are fair and equitable.

Continuous Improvement

Fairness is a dynamic concept that evolves. It is essential to continuously monitor and update the AI system to ensure it remains fair and aligned with our growing understanding of fairness.

In conclusion, measuring and evaluating fairness in AI systems is a multifaceted task that requires a holistic approach. It involves technical assessments and a deeper reflection on fairness's ethical, cultural, and philosophical dimensions. By approaching this task with rigor, sensitivity, and a commitment to continuous improvement, we can work toward developing AI systems that are truly fair, equitable, and beneficial for all.

Ensuring Accountability in AI Systems

Ensuring accountability in AI systems is a pivotal concern that traverses multiple dimensions of our existence, from the individual to the societal, from the technical to the philosophical. It is a journey that compels us to confront the fundamental principles that underpin our relationship with technology, authority, and each other.

Historical Perspective

The issue of accountability has been debated throughout human history, shaping our legal systems, ethical codes, and social contracts. The emergence of AI has reframed this age-old debate, introducing new challenges and opportunities. Historically, accountability has been attributed to humans or organizations that can be held responsible for their actions. However, with AI, the lines of responsibility become blurred as algorithms make decisions that their creators may not fully understand.

Philosophical Implications

From a philosophical standpoint, accountability is intrinsically linked to autonomy, agency, and responsibility. As AI systems gain more freedom and decision-making capabilities, questions arise about their agency and the extent to which they can be held accountable. Moreover, the philosophical underpinnings of accountability compel us to reflect on the nature of AI as an extension of human agency and the implications of this relationship for our ethical frameworks.

Legal Challenges

Legally, ensuring accountability in AI systems poses significant challenges. Existing legal frameworks are often ill-equipped to deal with the complexities of AI, as they are based on concepts of human agency and responsibility. Developing new legal frameworks that can address the specific challenges posed by AI is crucial for ensuring accountability.

Technical Considerations

On a technical level, ensuring accountability involves developing mechanisms to track and document the decisions made by AI systems and the data and algorithms that inform these decisions. This involves creating audit trails, implementing explainability features, and developing methods to reverse-engineer AI decisions when necessary.

Organizational Responsibility

Organizations that develop, deploy, or use AI systems are crucial in ensuring accountability. This involves implementing robust governance structures, developing ethical guidelines, and establishing procedures for monitoring and auditing AI systems. Moreover, organizations must be transparent about their use of AI and the potential impacts on stakeholders.

Public Engagement

Public engagement is essential for ensuring accountability in AI systems. This involves actively seeking input from diverse stakeholders, including the public, experts, and marginalized communities. By fostering a participatory approach to AI development and governance, we can ensure that the perspectives of all stakeholders are considered and that the benefits and risks of AI are distributed more equitably.

International Collaboration

AI systems often transcend national borders, which poses challenges for ensuring accountability. International collaboration is essential for developing global standards, sharing best practices, and coordinating efforts to regulate AI. This involves engaging with international organizations, governments, and civil society to create a global framework for AI accountability.

Ensuring accountability in AI systems is a herculean task that demands a multifaceted approach. It involves grappling with complex legal, technical, and ethical challenges and necessitates a commitment to continuous reflection and improvement. By embracing a holistic approach that encompasses historical, philosophical, legal, technical, organizational, and international dimensions, we can work toward creating AI systems that are accountable, transparent, and beneficial for all.

Real-World Examples

As the deployment of artificial intelligence systems becomes more widespread across various sectors of society, there is a growing need to ensure these systems are accountable, fair, and free from bias. Ensuring accountability in AI systems is a theoretical concept and a practical necessity. Several organizations, governments, and institutions have taken significant steps to address this issue and have implemented measures to ensure accountability in their AI systems. In this section, we will explore real-world examples of how different entities approach the issue of responsibility in AI systems, showcasing the diverse strategies and methods employed to address this critical concern.

Health Care AI Accountability

In health care, AI systems predict diseases, diagnose conditions, and recommend treatments. The US Food and Drug Administration (FDA) has been proactive in establishing a regulatory framework for AI in health care, including the requirement for a "Software as a Medical Device" (SaMD) to have an explicit algorithm change protocol. This means that any changes to the algorithm must be documented, validated, and submitted to the FDA for approval, ensuring accountability in the development and deployment of health care AI systems.

Self-Driving Car Accidents

Several incidents involving autonomous vehicles have raised questions about accountability in AI systems. For example, in 2018, a self-driving car operated by Uber struck and killed a pedestrian in Arizona. The incident prompted a thorough investigation and a temporary halt to Uber's self-driving car testing. The study revealed that the car's software detected the pedestrian but failed to act quickly. This incident highlighted the importance of clear accountability frameworks for autonomous vehicles and led to increased scrutiny and regulation of self-driving car testing.

AI in Judicial Systems

In the United States, an AI system called COMPAS (Correctional Offender Management Profiling for Alternative Sanctions) has been used to predict the likelihood of a defendant committing another crime. The system faced criticism when it was revealed that it showed bias against African American defendants. This prompted a discussion about the accountability of AI systems used in judicial decision-making and led to calls for increased transparency, oversight, and regulation of such systems.

AI in Hiring

Several companies have developed AI systems for screening job applicants. These systems analyze resumes, conduct video interviews, and assess candidates' skills. However, there have been instances where these systems have shown bias against certain groups of applicants. For example, a study found that an AI system used for hiring favored male candidates over female candidates. This led to increased scrutiny of AI systems used in hiring and called for greater accountability and transparency in their development and deployment.

Public Sector AI Accountability

In the United Kingdom, the government has established the Centre for Data Ethics and Innovation (CDEI) to advise on the governance of AI and data-driven technologies. The CDEI has developed a framework for public sector use of AI, which includes principles for transparency, accountability, and fairness. This is an example of a government taking proactive steps to ensure accountability in deploying AI systems in the public sector.

US Government Accountability Office

The US GAO developed the federal government's first framework to assure accountability and responsible use of AI systems. The framework defines the basic conditions for accountability throughout the entire AI life cycle, from design and development to deployment and monitoring. It lists specific questions to ask and audit procedures to use when assessing AI systems along four dimensions: governance, data, performance, and monitoring.

IBM

IBM has considered its responsibility when a hotel AI assistant's feedback does not meet the needs or expectations of guests. The company implemented a feedback learning loop to understand preferences better and highlighted the ability for a guest to turn off the AI at any point during their stay.

Google's AI Principles

Google published AI Principles that outline its commitment to developing and using AI responsibly. The principles include a commitment to transparency, fairness, and accountability.

The Partnership on AI

A collaboration between leading technology companies, academic institutions, and civil society organizations working to develop best practices for the responsible development and use of AI. The European Union's Artificial Intelligence Act

The European Union is developing an Artificial Intelligence Act to set requirements for creating and using AI systems. The act would require AI systems to be transparent, fair, and accountable.

The Algorithmic Accountability Act

A proposed law in the United States would require large technology companies to disclose how their algorithms work and to take steps to mitigate bias.

The AI Now Institute

The AI Now research institute studies the social implications of artificial intelligence. The institute is developing policies and practices to ensure that AI is used responsibly.

These examples collectively demonstrate the multifaceted efforts made across different sectors to ensure accountability in AI systems. From government frameworks to corporate principles and international collaborations, there is a concerted effort to ensure AI systems are transparent, fair, and accountable.

FUTURE CHALLENGES AND OPPORTUNITIES

As AI continues evolving and permeating every facet of human life, we must approach its management carefully, cautiously, and optimistically. Let us understand the challenges and opportunities that there are.

Challenges

Evolving Bias

Societies continuously evolve, and with them, so do biases. This fluidity means what might be considered a bias today could change tomorrow. Therefore, AI systems need constant updating and monitoring to prevent perpetuating outdated or newly emerged biases. This requires a dynamic and ongoing commitment to fairness and accountability beyond the initial design and implementation of AI systems.

Technological Advancements

As AI technology becomes more sophisticated, there is a risk that new forms of bias, which are not yet understood or accounted for, may emerge. For example, with the advent of more advanced natural language processing (NLP) techniques, there might be biases in how AI systems interpret or respond to different dialects or languages.

Globalization

Different cultures and societies have different perspectives on fairness and bias. What is considered fair or biased in one part of the world may not be perceived similarly in another. This presents a significant challenge in developing AI systems that are globally acceptable and do not inadvertently reinforce biases that are prevalent in one culture but not in another.

Legal and Ethical Dilemmas

Different regions have different legal and ethical standards. Balancing these considerations while developing and implementing AI systems can be complicated and challenging. For example, data privacy laws vary widely from region to region, and an AI system considered ethical and lawful in one country may not be so in another.

Opportunities

■ *Collaborative Efforts:* Collaboration between various stakeholders, including governments, academia, and the private sector, can lead to the development of more robust and fair AI systems. By pooling knowledge, resources, and perspectives, it is possible to develop AI systems that are more inclusive and better equipped to address and mitigate biases.

▪ *Innovative Solutions:* Technological advancements provide an opportunity to develop innovative solutions to address and mitigate bias in AI systems. For example, advanced machine learning techniques can identify and correct biases in the data used to train AI systems.

Public Awareness

As public awareness and understanding of the issues related to fairness, bias, and accountability in AI increase, there will be greater scrutiny and demand for responsible AI systems. This heightened awareness can lead to more informed decisions by consumers and policymakers alike, driving the development of fairer and more accountable AI systems.

Regulation and Standardization

Developing global standards and regulations can help ensure that AI systems are developed and implemented fairly and accountable. By creating a common framework that all stakeholders can adhere to, it is possible to create a level playing field and reduce the risk of unfair or biased AI systems being developed or deployed.

Addressing these challenges and capitalizing on these opportunities will ensure the responsible development and deployment of AI systems in the future.

TRANSPARENCY AND EXPLAINABILITY

Transparency and explainability are cornerstones of ethical artificial intelligence (AI). They ensure that the AI systems we create and implement operate fairly and justly. *Transparency* is sharing relevant information about the AI system with the involved stakeholders. It encompasses revealing the calculations that underpin decisions made by AI systems. Transparent AI systems share insights into their decision-making processes, shedding light on the reasoning behind their outputs.

Explainability concerns the accessibility, interpretation, and comprehension of the AI system's decision-making processes by relevant parties. It entails the AI software's capacity to articulate its conclusions and rationale to humans or other entities. Beyond merely providing accurate results, explainability offers insights into why specific decisions were made, thereby increasing the trustworthiness of AI systems and facilitating their oversight.

Together, these principles augment our understanding of AI decisions, facilitate examining AI outcomes, and promote accountability. By demystifying the obscurity of AI algorithms, we can confirm that the decision-making process is transparent, equitable, and in harmony with human values.

The imperative for transparency and explainability stems from several key reasons. First, they engender trust in AI systems by clarifying how they function and rationalize decisions. Second, they help uncover and rectify AI systems' biases, which are shaped by the data they are trained on. Finally, they support the responsible and ethical application of AI systems by enabling a thorough assessment of potential risks and benefits.

Despite these concepts' critical importance, significant obstacles exist to realize transparency and explainability in AI systems. The complexity of AI systems, the vast amounts of data they are trained on, and the absence of universally accepted metrics and assessments are all contributing factors. Nonetheless, there is increasing interest in and commitment to addressing these issues. Researchers, developers, governments, and organizations are working on novel techniques, guidelines, and regulations to advance transparency and explainability in AI.

As AI systems become more deeply embedded in our society, ensuring they operate with transparency and explainability is crucial. This will cultivate trust, minimize bias, and guarantee that AI systems are employed responsibly and ethically.

Explainable AI

Explainable AI is a field of artificial intelligence dedicated to elucidating the decision-making processes and operations of AI systems to humans. Traditional machine learning models and algorithms often lack transparency, functioning as "black boxes" that produce accurate predictions or results without revealing the underlying reasoning. Explainable AI seeks to overcome this limitation by delivering precise outcomes and elucidating the rationale behind specific decisions. This enhances the trustworthiness of AI systems and facilitates their management.

The need for explainability arises from several advantages:

- *User Trust:* Users are more likely to trust and follow the recommendations of a system that can explain its decisions and conclusions. This is particularly important in health care and finance, where decisions carry significant weight.
- *Decision-making Transparency:* Explainable AI elucidates the attributes or data that influence results, enabling the detection of potential issues or errors and the refinement of model training processes.
- *Bias Detection:* Explainable models help identify biases and discrimination that may emerge during learning, promoting fairer and more ethical AI systems.
- *Regulatory Compliance:* Certain sectors, like health care and finance, have stringent regulatory requirements for transparent decision-making. Explainable AI facilitates compliance with these mandates.

Explainable AI is employed in various domains:

- *Health Care:* It helps interpret signs and findings that lead to specific diagnoses or treatment recommendations, aiding physicians in making informed decisions and explaining them to patients.
- *Finance:* It assists in analyzing risks, determining customer creditworthiness, and detecting fraudulent activities, enabling banks and financial institutions to make informed decisions and minimize risks.
- *Autonomous Vehicles:* It offers insights into the data and sensors influencing vehicle decisions, enhancing passenger safety and usability.
- *Recommender Systems:* It can clarify why certain products or services were recommended to users, improving recommendation personalization and customer satisfaction.

Various methods and approaches exist for creating explainable models and algorithms:

- *Simple Models:* Utilizing simpler models, such as logistic regression or decision trees, can enhance the interpretability of artificial intelligence.
- *Interpretable Algorithms:* Special interpretable algorithms, such as LIME (Local Interpretable Model-agnostic Explanations) and SHAP (SHapley Additive exPlanations), generate models with clear rules and transparent logic.
- *Data Visualization:* Visualizing data helps understand the features or variables contributing most to the model's results.

Explainable AI is a critical advancement in artificial intelligence, making it more comprehensible, trustworthy, and transparent. It enables a deeper understanding of decision-making rationale and facilitates meaningful model and algorithm quality improvements. Applying explainable AI across various fields presents new opportunities for future artificial intelligence growth and development.

The Importance of Transparency and Explainability

In artificial intelligence, transparency is a necessity. As machines take on roles that were once performed by humans, the need for these machines to be transparent in their operations becomes paramount. If the inner workings of AI systems are unknown or unclear, we should be concerned because these mechanisms influence the decisions affecting our lives.

Transparency in AI systems involves showing the inner workings, algorithms, and data the system uses to decide. It is the antithesis of the "black box" approach, where the system's internal workings are hidden from view, and only the input and output are known. The "black box" approach is untenable in a world where AI systems increasingly influence critical decisions, from medical diagnoses to loan approvals.

The importance of transparency in AI systems can be distilled into three areas:

- *Trust:* Trust is important in any relationship, and it is no different regarding our relationship with AI systems. If we are to entrust machines with decisions that impact our lives, we need to have confidence in their decision-making processes. Transparency is critical to building this trust. When we understand how a system arrives at a decision, we are more likely to trust its judgment.
- *Accountability:* With great power comes great responsibility. As AI systems take on more influential roles, there needs to be a mechanism for holding them accountable for their decisions. Transparency enables accountability by allowing us to trace the decision-making process and identify any potential flaws or biases.
- *Ethical Decision-making:* Ethical considerations are becoming increasingly important as AI systems are deployed in more sensitive areas. Transparency allows us to scrutinize the decision-making process of AI systems to ensure that they align with our ethical values and societal norms.

Despite its importance, achieving transparency in AI systems is challenging. The complexity of the algorithms and the sheer volume of data makes it difficult to understand the decision-making process fully. Additionally, there is a trade-off between transparency and performance. Highly transparent models may not be as accurate as their opaque counterparts, leading to a dilemma between understandability and performance.

Nonetheless, the importance of transparency in AI systems cannot be overstated. It is fundamental to building trust, ensuring accountability, and aligning AI systems with our ethical values. As we move toward a future where AI systems play an even more prominent role in our lives, the quest for transparency will be more critical than ever.

Making AI Decisions Explainable

Explainability in AI is the capacity of a system to clarify and justify its decisions and actions in a way that humans can understand. It involves breaking down the complexities of an AI system's decision-making process into comprehensible pieces that humans can easily interpret and assess.

The drive for explainability stems from a fundamental need for comprehension, control, and accountability. As AI systems become more intricate and wield more influence over various aspects of our lives, from health care to finance to criminal justice, the ability to understand and interpret their decisions becomes imperative.

Here are some reasons why making AI decisions explainable is crucial:

- *Dynamic Trust Calibration:* Trust in AI systems should not be static; it should be dynamically calibrated based on the system's performance, context, and criticality of the decision. Dynamic trust calibration involves continuously assessing and adjusting the level of trust placed in an AI system based on real-time feedback and contextual factors.

- *Distributed Accountability:* In an interconnected world, AI systems often interact and influence each other. Accountability should not be solely attributed to a single AI system but spread across the ecosystem of interacting AI systems, developers, and users. This involves developing frameworks for assessing and attributing accountability in multi-agent AI systems and complex decision-making processes.
- *Continuous Ethical Monitoring:* Ethical compliance should not be a one-time assessment but a constant process. Continuous ethical monitoring involves real-time review of AI decisions against a dynamic set of moral principles and guidelines that evolve with societal norms and values.
- *Adaptive Decision-making:* AI systems should be capable of adapting their decision-making process based on the context and feedback. Adaptive decision-making involves developing AI systems that can learn from their mistakes, incorporate feedback, and adjust their decision-making process to improve decision quality over time.
- *Augmented Collaboration:* Human collaboration with AI should not be limited to understanding AI decisions. Increased collaboration involves designing AI systems that explain their decisions, actively seeking feedback, providing recommendations, and facilitating a two-way dialogue between humans and AI systems.

While the need for explainability is clear, achieving it is no small feat. AI systems and intense learning models are inherently complex and challenging to interpret. Moreover, there is a trade-off between explainability and model performance; simpler models that are easier to explain may not be as accurate as more complex models. Additionally, different stakeholders, such as end users, developers, and regulators, may require different levels of explainability.

Despite these challenges, efforts are being made to develop techniques and tools that can make AI decisions more explainable. These include the development of interpretable models, local explanation methods, and visualization tools. Additionally, there is a growing recognition of the importance of explainability in developing and deploying AI systems, as reflected in various guidelines and regulations that emphasize the need for explainability.

Making AI decisions explainable is crucial for building trust, ensuring accountability, and facilitating human collaboration. While it poses significant challenges, ongoing efforts in research and development are making strides toward achieving explainability in AI systems. Explainability will remain critical as AI continues to permeate our lives.

Challenges in Achieving Transparency and Explainability

Achieving transparency and explainability in AI systems is a multifaceted challenge that involves both technical and ethical dimensions. Here are some challenges.

Technical Challenges

Complexity of Models: Modern AI systems, especially deep learning models, are incredibly complex and involve thousands, if not millions, of parameters. This complexity makes it challenging to understand and interpret the decisions made by the AI system.

Trade-Off Between Performance and Explainability

There is often a trade-off between an AI system's performance (e.g., accuracy and speed) and its explainability. Highly accurate models may be less interpretable, whereas simpler, more interpretable models may not perform as well.

Lack of Standardized Explainability Metrics

There is a lack of standardized metrics and methods for assessing the explainability of AI systems. This makes it difficult to compare the explainability of different models and to develop universally accepted standards for explainability.

Ethical Challenges

Bias and Fairness

AI systems are trained on data that may contain biases. Even if the model itself is neutral, it may inadvertently learn and perpetuate these biases, leading to unfair or discriminatory decisions. Ensuring fairness and mitigating bias in AI systems is a significant ethical challenge.

Privacy Concerns

Explainability may require revealing sensitive information about the data or the decision-making process, which could raise privacy concerns. Striking the right balance between explainability and privacy is a critical ethical consideration.

Misuse of Explainability

There is a risk that explainability could be misused to provide a false sense of confidence in an AI system or to justify unethical decisions. For example, an AI system might offer a plausible but incorrect explanation for a decision to gain users' trust.

Addressing these challenges requires a multidisciplinary approach combining technical innovation and ethical considerations. It involves developing new methods and tools for explainability, creating standardized metrics for assessing explainability and developing ethical guidelines and regulations to ensure that explainability is used responsibly and ethically.

Techniques for Enhancing Transparency and Explainability

To enhance the transparency and explainability of AI systems, various methods and techniques are currently being developed or used.

Interpretable Models

Some models are inherently more interpretable, such as linear regression models, decision trees, and rule-based systems. These models can be used in situations where explainability is a high priority.

Model-Agnostic Methods

These methods can be applied to any machine learning model to make its decisions more interpretable. Examples include the following:

- *LIME (Local Interpretable Model-agnostic Explanations):* This method explains the predictions of any classifier by approximating it locally with an interpretable model.
- *SHAP (SHapley Additive exPlanations):* This method assigns each feature an importance value for a particular prediction based on a game theoretic approach.

Feature Selection and Reduction

Reducing the number of features a model uses can often make it more interpretable, as long as it does not significantly reduce its performance. Techniques such as Principal Component Analysis (PCA) or Recursive Feature Elimination (RFE) can be used for feature reduction.

Visualization Tools

Visual representations of data and model decisions can make them more interpretable. For example, visualizing a neural network's weights or a classifier's decision boundary can provide insights into its functioning.

Natural Language Explanations

Generating natural language explanations of model decisions can make them more accessible and understandable to non-technical users. For example, an AI system could generate a textual explanation of why it made a particular recommendation.

Counterfactual Explanations

These are explanations that describe the minimum change needed to the input features of a model to change its prediction. For example, a counterfactual explanation for a loan rejection might be: "The loan would have been approved if the applicant's income was $10,000 higher."

Explainable AI Platforms

Various platforms and toolkits, such as IBM's AI Explainability 360 or Google's What-If Tool, provide a range of techniques and visualizations to make machine learning models more interpretable.

It is important to note that there is no one-size-fits-all solution for enhancing transparency and explainability, as the appropriate method depends on the specific model, application, and requirements. Often, a combination of methods may be necessary to provide a comprehensive and understandable explanation of a model's decisions.

Real-World Examples of Transparent and Explainable AI

As the algorithms that power our world become increasingly complex, the importance of making AI systems transparent and explainable cannot be overstated. This is not just a technical necessity but a moral imperative. The following are some real-world examples of successful implementations.

IBM Watson in Health Care

IBM Watson is employed in various areas of health care, from diagnostics to treatment recommendations. Its AI algorithms can analyze the meaning and context of structured and unstructured data in clinical notes and reports, helping doctors make more informed decisions. Watson's decision-making process is designed to be transparent, allowing doctors to see all the information that led to a particular recommendation.

LIME (Local Interpretable Model-agnostic Explanations)

This is a popular open-source tool designed to help understand the predictions of any machine learning model. It works by approximating the model with a more interpretable one, locally around the prediction, making it easier for humans to understand the decision-making process.

FICO's Explainable Machine Learning Challenge

FICO, the company behind credit scores, hosted a challenge to create a machine learning model that could accurately predict credit risk while being explainable and transparent. The winners developed models that could accurately predict and clearly explain each prediction.

Google's Explainable AI (XAI)

Google Cloud has developed tools and frameworks to make their AI models more transparent and explainable. This includes visual tools that help users understand the model's feature importance, directional effects, and techniques for interpreting deep neural networks.

DARPA's Explainable Artificial Intelligence (XAI) Program

This program aims to create a suite of machine learning techniques that produce more explainable models while maintaining high learning performance. The goal is to enable human users to understand, trust, and manage the emerging generation of artificially intelligent partners.

These examples represent a small fraction of the ongoing efforts to make AI systems more transparent and explainable. As AI continues to evolve and play a more significant role in our lives, ensuring these systems are intelligent, understandable, and accountable will be crucial. The future of humanity may very well depend on our ability to make sense of the machines we create.

Future Prospects

The future trends and developments in transparency and explainability in AI systems are incredibly significant and multifaceted. In a world where the integration of artificial intelligence into daily life is becoming increasingly pervasive, several important trends and developments are expected to emerge.

Evolving Regulatory Landscape

Governments and regulatory bodies worldwide are becoming increasingly aware of the need for transparency and explainability in AI systems. As a result, new regulations and guidelines are expected to be implemented, mandating organizations to develop transparent AI systems that can be explained and justified. The European Union, for example, has already proposed regulations that require AI systems to be transparent and explainable.

Advanced Explainability Techniques

Researchers and practitioners are continuously working on developing more advanced techniques for making AI systems transparent and explainable. These techniques will likely involve a combination of different approaches, such as rule extraction, local interpretable model-agnostic explanations (LIME), and counterfactual explanations.

Integration of Human Expertise

Incorporating human expertise into developing and validating AI models will become more common. This will help design more transparent, explainable models more aligned with human values and ethical considerations.

Increased Use of Hybrid Models

Hybrid models combine the strengths of rule-based systems and machine-learning models and are expected to become more popular. These models can offer the best of both worlds: rule-based systems' interpretability and machine learning models' predictive power.

Development of Explainability Tools

As the demand for transparent and explainable AI systems increases, there will be a surge in the development of tools and platforms that facilitate the creation of such models. These tools will likely provide functionalities for visualizing model behavior, generating explanations, and assessing model fairness and robustness.

The development of artificial intelligence represents a monumental advancement in technology. However, as we continue to innovate, we must remain cognizant of these technologies' ethical implications and societal impacts. By prioritizing transparency and explainability in AI systems, we can ensure that these powerful tools are harnessed for the betterment of all rather than the detriment of some. As we forge ahead into this brave new world, let us do so with a spirit of collaboration, responsibility, and foresight.

DIGITAL DIVIDE AND INCLUSIVITY

Information is power, and the world is interconnected more than ever. There is, however, a *digital divide*, which refers to the gap between individuals, households, businesses, and geographic areas at different socioeconomic levels concerning their opportunities to access information and communication technologies (ICTs). This is a multifaceted issue that encompasses several dimensions.

- *Access to Technology:* The most basic level of the digital divide involves access to computers and the Internet. There is a gap between urban and rural areas and between developed and developing countries. A considerable portion of the world's population still does not have adequate access to the Internet or digital devices.
- *Digital Literacy:* Access alone is not sufficient. People need the skills to navigate the digital world effectively. This includes finding, evaluating, and creating information using digital technology. Unfortunately, many people, particularly older ones, lack these essential skills.

- *Quality of Content and Services:* Even with access and skills, the digital divide regarding the quality of content and services available still exists. Many regions, especially in developing countries, do not have access to high-quality educational content, government services, or health care information.
- *Affordability:* The cost of devices and Internet services can be a significant barrier for many people, particularly in low-income communities. While the cost of technology has decreased over time, it remains prohibitively expensive for many.
- *Language and Cultural Barriers:* A few major languages dominate the Internet, and much of the content is irrelevant or accessible to people who speak other languages. Additionally, cultural differences can make it difficult for people to understand or trust online information.

To build a more inclusive digital future, we must address all of these dimensions of the digital divide. This will require coordinated efforts from governments, the private sector, and civil society. It is not merely a matter of infrastructure and access but of education, content, affordability, and cultural relevance. Understanding the nature and nuances of the digital divide is the first step toward bridging it.

Impact of the Digital Divide

The digital divide, a term that refers to the disparities in access to, use of, and impact of ICT, has far-reaching consequences that permeate various aspects of society.

Education

The quality of education one receives is heavily influenced by access to digital resources. Online educational content, digital textbooks, and interactive learning platforms have become indispensable tools in modern education. Students without access to these resources are at a significant disadvantage. The divide is between developed and developing countries, within a single country, between urban and rural areas, and between different socioeconomic groups. This leads to a widening educational gap, limiting future opportunities for those on the wrong side of the divide.

Health Care

Digital technology is crucial in modern health care, from telemedicine to online health information and appointment booking systems. The digital divide affects access to these services and the quality of health care one receives. For example, in remote areas where Internet access is limited, telemedicine, which has proven to be a lifesaver during the pandemic, is virtually nonexistent. This can lead to delayed diagnoses, inadequate treatment, and poorer health outcomes.

Economic Opportunities

The Internet has revolutionized how we work, shop, and conduct business. Job opportunities, particularly for remote work, are abundant online, and e-commerce has opened up new avenues for businesses to reach customers worldwide. However, for those without access to the Internet, these opportunities remain out of reach. This leads to a vicious cycle of poverty, as lack of access to economic opportunities further exacerbates inequality.

Political Participation

The Internet is an important platform for political participation, from accessing news and information to engaging in online debates and voting in some countries. However, the digital divide means that a significant portion of the population is excluded from these activities. This can lead to a lack of political awareness, reduced civic participation, and a less robust democracy.

The digital divide is not just a technological issue but a social justice issue. It exacerbates existing inequalities and creates new ones. It hinders social mobility, economic development, and democratic participation. Addressing the digital divide is not merely a matter of improving access to technology but a matter of creating a more equitable and inclusive society.

Role of AI in Bridging the Digital Divide

AI has the potential to play a pivotal role in bridging the digital divide by enabling personalized services, improving access to information, and optimizing the delivery of essential services.

Personalized Education

AI can help bridge the educational divide by enabling personalized learning experiences. AI-powered educational platforms can analyze students' learning styles, strengths, and weaknesses to provide customized learning paths, resources, and feedback. This can be particularly beneficial for students in remote or underserved areas who may not have access to high-quality teaching resources. Moreover, AI can help translate language and create content in multiple languages, making education accessible to a broader audience.

Telemedicine

AI-powered telemedicine can help bridge the health care divide by enabling remote diagnostics and treatment. AI algorithms can analyze medical images, monitor patient data, and provide preliminary diagnoses, reducing the need for specialized medical personnel in remote areas. Moreover, AI-powered chatbots can provide basic health information and guidance, helping to address the shortage of health care professionals in underserved areas.

Better Access to Information and Services

AI can help improve access to information and services by enabling intelligent search and recommendation systems. These systems can help users find relevant information, products, or services more efficiently, reducing the barriers to access. Moreover, AI-powered voice assistants and chatbots can make it easier for people with low literacy levels or disabilities to access information and services.

Optimizing Infrastructure

AI can help optimize the delivery of essential services, such as electricity, water, and transportation, by predicting demand, optimizing supply, and improving maintenance. This can help reduce the infrastructure gap in underserved areas, enhancing the quality of life for their inhabitants.

Challenges in Promoting Inclusivity in AI

Promoting inclusivity in AI ensures that the technology benefits a wide range of people, regardless of their background, abilities, or socioeconomic status. However, several challenges need to be addressed to achieve this goal.

Biases in Data and Algorithms

AI systems are trained on large datasets, and if these datasets contain biases, the AI systems will inadvertently reproduce and possibly amplify these biases. For example, a facial recognition system trained predominantly on images of light-skinned individuals may have difficulty accurately recognizing dark-skinned faces. Similarly, an AI system for screening job applicants may unfairly favor specific demographics over others if the training data contains biases. Addressing these biases requires a concerted effort to curate diverse and representative datasets and to develop algorithms that are robust to biases.

Lack of Representation in AI Development Teams

The development of AI systems is often dominated by individuals from specific demographics, leading to a lack of diverse perspectives in designing and implementing these systems. This can result in AI systems that are optimized for certain groups of people but may be suboptimal or even harmful to others. Promoting diversity in AI development teams ensures that the technology is inclusive and beneficial for a broad range of people.

Need for Inclusive Design Principles

Inclusive design principles, which aim to create products and services that are accessible and usable by as many people as possible, are not always followed in the development of AI systems. For example, voice-activated AI systems may not work well for individuals with speech impairments, and visual interfaces may be inaccessible to visually impaired individuals. Incorporating inclusive design principles from the outset is essential to ensure that AI systems are accessible and usable by a wide range of people.

Economic and Geographic Inequalities

Economic and geographic inequalities can result in a lack of access to AI technologies for certain groups of people. For example, rural areas may lack the necessary infrastructure to support AI technologies, and economically disadvantaged individuals may be unable to afford devices or services that leverage AI. Addressing these inequalities requires a multifaceted approach that includes investments in infrastructure, affordable access to devices and services, and targeted interventions to support marginalized communities.

Ethical and Legal Considerations

Ethical and legal considerations, such as privacy, consent, and data protection, are critical in promoting inclusivity in AI. Ensuring that AI systems respect individuals' rights and values is essential to building trust and acceptance among users.

Addressing these challenges requires coordinating efforts among various stakeholders, including governments, industry, academia, and civil society. Moreover, involving the target populations in designing and implementing AI solutions is vital to ensure they are culturally appropriate and address their specific needs and preferences.

Best Practices for Inclusive AI

Inclusive AI is a critical aspect of ensuring that technology serves the needs of a diverse global population. However, creating inclusive AI systems involves overcoming several barriers, including biases in data, lack of representation in development teams, and economic and geographic inequalities. To address these challenges, here are some best practices for developing and implementing inclusive AI systems.

Diverse Data Collection

Collect and curate a diverse and representative dataset to train AI systems. This includes data from different demographics, geographies, and socioeconomic backgrounds to ensure the system can understand and serve a broad range of users.

Bias Detection and Mitigation

Implement methods to detect and mitigate biases in the data and algorithms. This involves using statistical methods to identify potential biases and developing algorithms that are robust to these biases.

Diverse Development Teams

Assemble diverse development teams that include individuals from various backgrounds, disciplines, and perspectives. This ensures that the AI system is designed and developed with a broad range of users in mind.

Inclusive Design Principles

Adhere to inclusive design principles, which aim to create products and services that are accessible and usable by as many people as possible. This includes designing interfaces that are accessible to individuals with disabilities and providing multilingual support.

User-Centric Development

Involve the target users in the design and development process to ensure that the AI system addresses their needs and preferences. This includes conducting user research, usability testing, and soliciting feedback from the target population.

Ethical and Legal Considerations

Ensure the AI system adheres to ethical principles and legal regulations like privacy, consent, and data protection. This involves conducting ethical assessments, obtaining necessary approvals, and implementing appropriate safeguards.

Affordable Access

Develop strategies to provide affordable access to AI systems for economically disadvantaged individuals and communities. This includes creating cost-effective solutions, leveraging existing infrastructure, and collaborating with local organizations.

Continuous Monitoring and Improvement

Implement mechanisms for continuous monitoring and improvement of the AI system. This includes collecting user feedback, analyzing the system's performance, and making necessary adjustments to ensure that it remains inclusive and effective.

By adhering to these best practices, it is possible to develop and implement AI systems that are inclusive, accessible, and beneficial to a broad range of people. It is essential to recognize that inclusivity in AI is not a one-time effort but an ongoing commitment that requires continuous attention and action.

Case Studies

The journey toward a more inclusive and equitable digital future is filled with challenges and triumphs. It is essential to acknowledge the hurdles that lie ahead and recognize and learn from the successes that have been achieved thus far. Let's consider some of them.

Case Study 1: Project Loon

Overview: Project Loon was an initiative by Alphabet, Google's parent company, to provide Internet access to remote and rural areas worldwide using high-altitude balloons. Although the project was officially closed in 2021, it provided valuable insights into innovative ways to bridge the digital divide.

Impact: After Hurricane Maria, Project Loon successfully provided Internet access to several remote regions worldwide, including Puerto Rico and parts of Kenya. It demonstrated the potential of using innovative technologies to bridge the digital divide and provide Internet access to underserved communities.

Lessons Learned: While the project faced several challenges, including technical difficulties and regulatory hurdles, it highlighted the importance of collaboration between governments, the private sector, and local communities in addressing the digital divide. It also underscored the need for sustainable and cost-effective solutions.

Case Study 2: Microsoft's AI for Accessibility

Overview: Microsoft's AI for Accessibility is a grant program that provides funding, technology, and expertise to organizations working on AI-powered solutions to improve the lives of people with disabilities. The program focuses on employment, daily life, and communication and connection.

Impact: The program has funded several innovative projects, including an AI-powered app that helps people with visual impairments identify objects and navigate their environment and a virtual reality platform that helps people with mobility challenges participate in physical therapy at home.

Lessons Learned: The program highlights the importance of leveraging AI technology to create solutions that address the unique needs of people with disabilities. It also underscores the importance of collaboration between tech companies, non-profit organizations, and the disability community in developing inclusive AI solutions.

Case Study 3: AI4ALL

Overview: AI4ALL is a non-profit organization that aims to increase diversity and inclusion in artificial intelligence. The organization runs education programs for underrepresented high school students, provides mentorship and support for college students and early-career professionals, and researches diversity and inclusion in AI.

Impact: AI4ALL has reached thousands of students from underrepresented backgrounds, helping them gain the skills and confidence to pursue careers in AI. The organization's alumni have gone on to win prestigious scholarships, secure internships at leading tech companies, and conduct research at top universities.

Lessons Learned: AI4ALL demonstrates the importance of providing education, mentorship, and support to underrepresented individuals to promote diversity and inclusion in AI. It also highlights the need for a multifaceted approach addressing the various barriers underrepresented individuals face in accessing and succeeding in AI.

Case Study 4: Bridging the Digital Divide with Starlink

Overview: Starlink is an ambitious initiative SpaceX developed to provide global Internet coverage through a constellation of thousands of low-orbit satellites. With over 2,000 satellites already launched and commercial access offered in select locations, Starlink promises to provide Internet speeds of up to 1 Gbps, revolutionizing Internet access in developing countries and remote areas.

Impact: The potential effects of Starlink are enormous, as it could transform education, health care, and economic opportunities for millions worldwide who currently have limited or no Internet access. The project has faced criticism for its environmental impact and lack of transparency, raising concerns about the potential negative consequences of launching thousands of satellites into orbit.

Lessons Learned: While Starlink offers a promising solution to bridging the digital divide, it highlights the importance of addressing potential environmental and social costs. As the project progresses, it will be crucial for SpaceX and other stakeholders to carefully weigh the benefits against the potential negative impacts and work toward solutions that maximize the positive effects while minimizing the negative consequences.

AUTONOMY AND DECISION-MAKING

With the increasing capabilities of AI systems, there is a growing reliance on these technologies to automate and optimize decision-making processes. This shift brings with it a host of challenges and opportunities that require careful consideration and strategic planning.

Autonomy in this context refers to the capacity of AI systems to make decisions and perform tasks without human intervention. This autonomy is crucial for various applications, from autonomous vehicles to automated trading systems and intelligent personal assistants.

However, the increasing autonomy of AI systems raises several questions. How can we ensure that these systems make decisions that are ethical, fair, and aligned with human values? How can we balance the convenience and efficiency of automation with the need to preserve human autonomy and agency? What are the potential risks and challenges associated with delegating decision-making to AI systems, and how can we mitigate them?

These are just a few of the many questions this section addresses. We will explore the key concepts and challenges related to AI autonomy and decision-making, including the different models of human-AI interaction, the role of AI in overcoming human cognitive biases, and the ethical considerations associated with AI decision-making. We will also examine the impact of AI on organizational structures, human skills, and governance and discuss strategies for designing AI systems that support rather than undermine human autonomy and decision-making.

Definitions and Difference

In a world increasingly reliant on technology, two terms often surface in discussions about innovation and security: "autonomous" and "automated." These seemingly similar concepts play distinct roles in artificial intelligence and are crucial to understanding the landscape of modern security strategies.

Picture a bustling, futuristic city. The air hums with drones flying by, delivering packages to high-rise apartments. On the streets below, self-driving cars precisely navigate the roads. In a nearby office building, a network of computers is hard at work, monitoring the city's infrastructure for any signs of trouble. This scene embodies the essence of autonomy and automation, two concepts underpinning our modern world's functioning.

Autonomous Systems as "Sentient Beings"

In artificial intelligence, *autonomy* refers to the ability of a system to make decisions and perform tasks without human intervention. Imagine a self-driving car navigating a busy street. It must continuously make decisions about speed, direction, and whether to stop or go, all in real time. These decisions are made by the car's onboard computer, which processes data from its sensors and makes decisions based on its programming and learned experiences.

Autonomous systems are designed to learn from their environment and adapt to new situations. They can analyze vast amounts of data, identify patterns, and make decisions based on that analysis. This ability to learn and adapt makes autonomous systems incredibly powerful and versatile.

Automated Systems as the "Workhorses of the Modern World"

Automation refers to using machines or computers to perform tasks that humans previously did. Automated systems follow pre-programmed rules and cannot learn or adapt to new situations. For example, a factory assembly line that repetitively performs the same task, day in and day out, is an example of automation.

Automated systems are designed to perform specific tasks with high efficiency and consistency. They can handle repetitive tasks that would be tedious or impractical for humans. Automation can increase productivity, lower costs, and provide higher-quality products or services.

Designing Security Strategies

Understanding the differences between automation and autonomy is crucial when designing security strategies. Automated systems can be incredibly efficient at performing repetitive tasks but may be vulnerable to unexpected situations or malicious attacks. An automatic security system, for example, may be programmed to detect certain types of cyberattacks and respond in a predetermined way. However, if it encounters a new attack that it has not been programmed to recognize, it may fail to react adequately.

An autonomous security system can learn from past experiences and adapt to new threats. It can analyze vast amounts of data in real time, identify patterns of malicious activity, and respond in a way that neutralizes the threat. This ability to learn and adapt makes autonomous systems more resilient to new and evolving threats.

Autonomy also raises ethical and accountability concerns. If an autonomous system makes a decision that leads to harm or loss, who is responsible? The designers of the system? The operators? These questions need to be carefully considered when designing and implementing autonomous systems.

Understanding the nuances between autonomous and automated systems is crucial for designing effective security strategies. While automated systems excel at performing repetitive tasks with high efficiency, they may struggle with unexpected situations. Autonomous systems can learn and adapt to new threats, making them more resilient to evolving challenges. Their use also raises ethical and accountability questions that must be carefully considered. Ultimately, a well-designed security strategy may involve a combination of both automated and autonomous systems, leveraging the strengths of each to create a robust and resilient defense.

Human-AI Interaction Models

In the not-too-distant future, the lines between humans and machines may become increasingly blurred. Artificial intelligence will likely become an integral part of our daily lives, assisting us in everything from driving our cars to managing our finances. As AI systems become more advanced and integrated into our lives, it is essential to establish effective interaction models between humans and machines. Four main models have emerged: Human-in-the-Loop (HITL), Human-in-the-Loop with Final Edit (HITLFE), Human-on-the-Loop (HOTL), and Human-out-of-the-Loop (HOOTL). Each of these models has advantages and disadvantages, and the choice of which one to adopt depends on the specific use case and desired level of human involvement.

The Meeting Room Dilemma

Imagine a scenario in a corporate meeting room where a group of executives is gathered to discuss the company's strategic direction. The meeting is facilitated by an AI-powered assistant who helps manage the agenda, take notes, and provide relevant data and insights to aid decision-making.

In a HITL scenario, the AI assistant actively participates in the meeting by providing real-time analysis and suggestions. However, the final decisions are made by the human executives. This model allows for a collaborative approach, leveraging the strengths of both humans and AI. The AI assistant can process vast amounts of data quickly and provide insightful analysis, while the human executives can apply their judgment, experience, and intuition to make the final decisions. This model suits situations where the stakes are high, and human review is essential.

In a HITLFE scenario, the AI assistant provides a draft of the meeting minutes, decisions, and action items at the end of the meeting. A human executive reviews and edits the final document before it is circulated. This model provides a balance between efficiency and accuracy. The AI assistant can quickly generate a draft document, but the human executive ensures it is accurate and complete.

In a HOTL scenario, the AI assistant manages the entire meeting autonomously, from setting the agenda to documenting the decisions and action items. The human executives are present and can intervene if necessary, but the AI assistant is designed to handle most tasks independently. This model suits routine meetings where the decisions are relatively straightforward and do not require high human judgment.

In a HOOTL scenario, the AI assistant manages the entire meeting without human intervention. The human executives receive a summary of the meeting and the decisions made, but they do not participate in the meeting itself. This model is suitable for situations where the decisions are highly data-driven and do not require human judgment or intuition.

The Pros and Cons

Each of these models has its own set of advantages and disadvantages. HITL and HITLFE models provide a high level of human control and judgment, which may be necessary for critical decisions or complex situations. They may also be less efficient, as they require human intervention and review. HOTL and HOOTL models are more efficient as they reduce the need for human intervention. They may be less suitable for situations that require human judgment, intuition, or ethical considerations.

Guidance for Selecting the Appropriate Model

The choice of which model to adopt depends on several factors.

- *The complexity of the task:* Complex tasks that require human judgment, intuition, or ethical considerations may be better suited for HITL or HITLFE models. Routine tasks that are highly data-driven may be more suitable for HOTL or HOOTL models.
- *The level of trust in the AI system:* If the AI system has been thoroughly tested and proven reliable, it may be appropriate to adopt a HOTL or HOOTL model. If there is less trust in the AI system, or if the stakes are high, assuming a HITL or HITLFE model may be safer.
- *The desired level of human control:* Some organizations may prefer to maintain a high level of human control over decision-making, while others may be more comfortable delegating decisions to AI systems.

Choosing which model to adopt is a strategic decision that depends on the organization's specific needs, goals, and level of trust in the AI system. It is essential to carefully consider these factors and choose the model that best aligns with the organization's objectives.

The Evolution of Decision-Making

Decision-making used to be an art that relied heavily on human judgment. Business leaders, government officials, and everyday people made decisions based on their experiences, intuition, and "gut feelings." It was a time when cognitive biases, often unbeknownst to the decision-makers themselves, played a significant role in the choices made.

The Influence of Cognitive Biases

Cognitive biases, such as confirmation bias, the availability heuristic, and anchoring effect, subtly influence decisions in ways that people often do not realize. *Confirmation bias* leads people to favor information that confirms their preexisting beliefs. The *availability heuristic* causes them to rely on readily available information rather than seeking out all relevant data. The *anchoring effect* influences people to rely heavily on the first piece of information they encounter, even if it was not the most important or relevant. These biases often lead to suboptimal decisions, but human judgment has been the best tool available in a world where data can be scarce and difficult to access.

The Rise of Data-Driven Decision-Making

As technology advanced and data became more accessible, the decision-making process evolved. Businesses started to collect and analyze data to inform their decisions, leading to a shift from purely intuitive decision-making to data-supported decision-making. This shift significantly improved decision quality as businesses could make more informed choices based on objective data rather than subjective opinions.

This new approach also brought its own set of challenges. The sheer volume of data available made it difficult for humans to process and analyze it all. Additionally, the cognitive biases that had influenced decisions in the past did not disappear overnight. Even with data at their fingertips, people still struggle to overcome these biases and make objective decisions.

The Shift to AI-Driven Workflows

As the limitations of human data processing became more apparent, businesses started to turn to artificial intelligence for help. With their ability to process vast amounts of data quickly and identify patterns humans might miss, AI algorithms have become an invaluable tool in decision-making. This shift from data-driven to AI-driven workflows marked a significant milestone in the evolution of decision-making.

AI not only helped businesses process data more efficiently, but it also helped overcome some of the cognitive biases that had plagued decision-making in the past. By providing objective, data driven insights, AI helped reduce the influence of prejudices such as confirmation bias and availability heuristics.

The Limitations of Human Processing

Despite the advancements in AI, human involvement in the decision-making process is still essential. Humans bring a level of creativity, empathy, and ethical considerations that AI systems currently lack. The limitations of human processing, such as the inability to process vast amounts of data quickly and the susceptibility to cognitive biases, remain a challenge.

As we move forward into a world where AI and human decision-making coexist, it is crucial to acknowledge and address these limitations. By leveraging AI's and humans' strengths, we can create a more efficient and effective decision-making process that leads to better outcomes for all.

The Journey Continues

The evolution of decision-making, from human judgment to data-supported decision-making and AI-driven workflows, has been marked by challenges and triumphs. We must remain adaptable and open to new approaches as we navigate this evolving landscape. By combining the best of what both humans and

AI have to offer, we can create a decision-making process that is more efficient, effective, and, ultimately, more beneficial for everyone.

The Role of AI in Decision-Making

As discussed earlier, decision-making has grown increasingly complex in a world inundated with data and information. While remarkable in many ways, humans are subject to cognitive biases that can impair judgment and lead to less-than-optimal decisions. Artificial intelligence is a technology capable of overcoming some of these biases by processing data at fine-grain levels, dealing with nonlinear relationships, and integrating seamlessly with human processors in workflows.

Overcoming Human Cognitive Biases

Imagine a scenario where Sarah, a seasoned manager at a tech company, is tasked with selecting a team for a new project. Without realizing it, Sarah may be influenced by various cognitive biases. For example, she might favor team members she has worked with before (availability bias) or give more weight to recent performance rather than considering a longer track record (recency bias).

AI can help overcome these biases by objectively analyzing potential team members based on past performance, skills, and availability. By considering a broader set of data and minimizing the influence of cognitive biases, AI can help Sarah make a more informed and objective decision.

Processing Data at Fine-Grain Levels

In another part of the world, David, a supply chain manager at a manufacturing company, is struggling to optimize inventory levels. The task involves analyzing vast amounts of data, including historical sales data, current inventory levels, and supplier lead times. It is a daunting task requiring data processing at fine-grain levels to make accurate predictions and decisions.

AI algorithms, capable of processing vast amounts of data quickly and accurately, can provide David with insights that would be almost impossible to derive manually. For example, AI can identify patterns in historical sales data that can help predict future demand with higher accuracy. By processing data at fine-grain levels, AI can help David make better-informed decisions and optimize inventory levels more effectively.

Dealing with Nonlinear Relationships

Emma, a financial analyst at a bank, is trying to create a model to predict loan defaults. The relationships between various factors, such as income level, credit history, and economic conditions, are complex and nonlinear. Traditional statistical methods struggle to capture these relationships accurately.

AI algorithms, particularly neural networks, are well-suited to model these complex, nonlinear relationships. By analyzing vast amounts of data and identifying patterns that might not be apparent to a human analyst, AI can help Emma create a more accurate and robust model for predicting loan defaults.

Integrating AI and Human Processors in Workflows

In all these scenarios, it is clear that AI is not a replacement for human judgment but rather a tool to enhance it. Sarah, David, and Emma all bring unique skills and knowledge that AI cannot replicate. However, by leveraging AI's and human processors' strengths, they can make better-informed decisions and achieve better outcomes.

AI is crucial in decision-making by overcoming human cognitive biases, processing data at fine-grain levels, dealing with nonlinear relationships, and integrating seamlessly with human processors in workflows. By leveraging AI's and humans' strengths, we can create more efficient and effective decision-making processes that lead to better outcomes for all.

Implementing AI in Organizational Structure

Let's consider another situation. Emma, the Chief Technology Officer (CTO), was about to lead a crucial meeting on implementing artificial intelligence into the company's organizational structure. The top management knew that integrating AI was no longer a matter of choice but a necessity to remain competitive. However, they also recognized that this integration needed to be thoughtful and strategic to ensure human managers' responsibility for AI output and foster human-machine collaboration.

Emma opened the meeting with a powerful statement, "AI is not here to replace us, but to enhance our capabilities. We aim to create a symbiotic relationship between humans and machines, where each complements the other's strengths and compensates for their weaknesses."

Integrating AI into Organizational Structure and Process Design

Emma explained that the first step in implementing AI into the organizational structure was to assess the current processes and identify areas where AI could add the most value. She used the example of the customer service department, which was overwhelmed with a high volume of inquiries. The company could handle routine inquiries more efficiently by implementing AI-powered chatbots, freeing human agents to deal with more complex issues requiring empathy and critical thinking.

She also emphasized the importance of redesigning processes to accommodate the integration of AI. "Simply adding AI to existing processes will not yield the desired results. We need to rethink our processes from the ground up to ensure that AI and human capabilities are leveraged optimally."

Ensuring Human Managers' Responsibility for AI Output

Emma then addressed a concern on everyone's mind: accountability for AI output. "While AI can process vast amounts of data and provide recommendations, the human manager should always make the final decision," she said. "This not only ensures accountability but also leverages the unique strengths of humans, such as judgment, intuition, and the ability to consider ethical implications."

She highlighted the importance of training managers to understand the AI algorithms' workings and limitations to make informed decisions. "We cannot blindly trust AI recommendations. Managers must have the knowledge and skills to assess these recommendations and make the final decision critically."

Fostering Human-Machine Collaboration

Finally, Emma discussed the importance of fostering a culture of collaboration between humans and machines. "We need to move away from the mindset of humans versus machines and toward a mindset of humans and machines working together," she said.

She shared an example of a project where a team of data scientists, engineers, and a machine learning algorithm worked together to develop a new product. "Each team member, human or machine, brought unique skills and perspectives. By leveraging these diverse capabilities, we were able to create a product that was superior to what any human or machine could have created alone."

Emma concluded the meeting by reiterating the importance of thoughtful and strategic implementation of AI into the organizational structure. "AI can potentially transform our organization, but only if we approach it thoughtfully and strategically. Let us work together to create a future where humans and machines collaborate to achieve greatness."

As the members left the meeting, there was a sense of optimism. They understood the challenges ahead but were excited about the possibilities AI could bring to their organization. With a clear vision and a strategic approach, they were ready to implement AI into their organizational structure.

Measuring Decision-Making Effectiveness

At the headquarters of a leading e-commerce company, the data analytics team was engrossed in a mission-critical task: measuring the effectiveness of the decision-making process. Sarah, the head of the team, understood the gravity of the situation. The company had recently implemented an AI-powered decision-making tool, and it was imperative to monitor its performance to ensure it was delivering the desired results.

Sarah gathered her team and outlined the plan. "We need to track key metrics that will help us assess the decision-making process's effectiveness. These metrics will help us understand how well the tool performs and identify areas where we need to make adjustments."

The team decided to track the following key metrics:

- *decision accuracy:* the percentage of correct decisions made by the AI tool
- *decision speed:* the time it took for the AI tool to decide
- *decision impact:* the overall impact of the decisions made by the AI tool on the business outcomes

As the team monitored these metrics over time, they noticed a trend. The decision accuracy and speed were consistently high, but the impact varied. This indicated that while the AI tool was making accurate and quick decisions, the overall effect of these decisions on the business outcomes was not as expected.

Sarah called for a meeting to discuss the findings and make necessary adjustments. "While the AI tool is performing well in accuracy and speed, we need to dig deeper to understand why the decision impact is not as expected," she said.

After a thorough analysis, the team discovered that the AI tool did not consider certain external factors influencing business outcomes. Sarah and her team worked diligently to incorporate these factors into the AI tool's decision-making process.

Over time, the decision's impact improved, and the AI tool became an invaluable asset to the company. Sarah and her team learned a valuable lesson: monitoring the decision-making process and making necessary adjustments is critical to ensuring the AI tool's success.

Choosing the Right System

Is choosing the right system crucial for an organization's success? John, the CTO of a cybersecurity firm, was faced with a daunting task: deciding between implementing an automated or an autonomous system for network intrusion detection.

John convened a meeting with his team to discuss the options. "An automated system will follow predefined rules and algorithms to detect network intrusions, while an autonomous system will use AI and machine learning to learn from past data and make decisions on its own," he explained.

The team discussed the potential benefits and drawbacks of each approach. An automated system would be easier to implement and manage, but it might not be as effective in detecting new types of network intrusions. While an autonomous system would be more capable of detecting new threats, it would be more complex to implement and manage.

After a thorough discussion, John and his team implemented an autonomous system. They believed the benefits of detecting new threats and adapting to changing circumstances outweighed the challenges of implementing and managing the system.

John and his team worked diligently to implement and train the autonomous system with past data. Over time, the system became more efficient and effective in detecting network intrusions, proving that choosing the right system was crucial for the firm's success.

Use Cases

In cybersecurity, network intrusion detection is important. Cybersecurity firm CyberGuard recently implemented an autonomous system for network intrusion detection and infrastructure compliance checks.

The system used machine learning algorithms to analyze network traffic and identify patterns indicative of a cyberattack. It also monitored the organization's infrastructure to ensure compliance with security policies and regulations.

Within the first month of implementation, the system detected a sophisticated cyberattack that had gone unnoticed by the firm's previous automated system. The autonomous system detected the attack, immediately isolated the affected network segments, and prevented the attack from spreading.

In another instance, the system identified a noncompliance issue in the organization's infrastructure that could have led to a security breach. The system alerted the IT team, who promptly addressed the issue and prevented a potential cyberattack.

These real-world examples highlight the capabilities of automated and autonomous systems in cybersecurity, network intrusion detection, and infrastructure compliance checks.

Impact of AI/ML

AI and machine learning have profoundly impacted the enhancement of automated and autonomous systems' capabilities. At TechSolutions, a leading tech firm, implementing AI and machine learning in their automated systems significantly improved efficiency, resilience, and decision-making capabilities.

The firm's automated systems, powered by AI and machine learning, could process vast amounts of data, identify patterns, and make decisions in real-time. This led to faster and more accurate decision-making, enabling the firm to respond to changing circumstances more effectively.

Furthermore, AI and machine learning algorithms enabled automated systems to learn from past data and improve their decision-making capabilities over time. This led to more resilient and efficient systems adapting to new challenges and making more sophisticated decisions.

Integrating AI and machine learning into automated and autonomous systems at "TechSolutions" led to more resilient, efficient, and capable systems that contributed to the firm's success.

Challenges and Risks

Implementing AI-powered decision-making tools is not without its challenges and risks. At DataCorp, a data analytics firm, implementing an AI-powered decision-making tool led to several challenges.

First, there was the risk of over-automation. The tool was designed to automate most of the decision-making process, but the management quickly realized that some decisions still required human intervention. They had to find the right balance between automation and human involvement to ensure optimal decision-making.

Second, there was the challenge of the black-box nature of some algorithms. It was difficult for the management to understand how the tool was making certain decisions, leading to a lack of trust and confidence in the tool.

Finally, there was difficulty in locating the cause of poor decision-making. When the tool made a suboptimal decision, it was challenging to identify the root cause and make necessary adjustments.

To address these challenges, DataCorp implemented a robust monitoring and evaluation system to track the AI tool's performance and make necessary adjustments. They also trained the management team to better understand the tool's workings and limitations.

Over time, DataCorp overcame these challenges and successfully implemented the AI-powered decision-making tool. However, the experience highlighted the importance of being aware of the potential challenges and risks associated with implementing AI-powered decision-making tools and having a plan to address them.

Ethical Considerations in AI Decision-Making

AI systems are increasingly used for decision-making, and ethical dilemmas are bound to arise. At EthicalAI, a company specializing in ethical AI solutions, the team grappled with some of these dilemmas.

One major concern was biased algorithms. The AI systems were trained on historical data, which often carried biases from the past. This resulted in biased decisions that unfairly affected certain groups of people. Another concern was privacy. The AI systems needed access to vast amounts of data to make informed decisions, but this raised concerns about individuals' privacy.

Finally, there was the issue of accountability. Who would be held accountable if an AI system made a decision that led to negative consequences? The AI system, the developers who created it, or the organization that implemented it?

At EthicalAI, the team worked diligently to address these ethical dilemmas. They implemented measures to identify and correct biases in the algorithms, implemented strict privacy safeguards, and developed a framework for accountability in AI decision-making.

Corporate Interests and AI Decision-Making

Corporate interests play a significant role in shaping AI decision-making processes, which can affect individual autonomy and societal well-being. At CorpTech, a leading tech firm, the management was pressured by shareholders to maximize profits. This led to the implementation of an AI decision-making system designed to optimize profitability, sometimes at the expense of other essential factors, such as employee well-being or environmental sustainability.

This raised concerns about the potential implications for individual autonomy and societal well-being. While the AI system successfully maximized profits, it also led to increased workloads for employees, higher levels of stress, and a negative environmental impact.

The experience at CorpTech highlights the importance of considering corporate interests' role in shaping AI decision-making processes and the potential implications for individual autonomy and societal well-being.

Balancing Convenience and Autonomy

AI systems offer unprecedented levels of convenience, but there is a trade-off between convenience and autonomy. At ConvenienceAI, a company specializing in AI-powered convenience solutions, the team was working on an AI system that could make decisions on behalf of users, such as ordering groceries, scheduling appointments, and managing finances.

While the system offered unparalleled convenience, it also raised concerns about the erosion of human autonomy. If the AI system made all the decisions, what role was left for the individual?

To address this concern, the team at ConvenienceAI implemented features that allowed users to set their preferences and have the final say on important decisions. This helped balance the benefits of AI-powered convenience with preserving human autonomy.

Designing AI for Human Autonomy

Designing AI systems that support rather than undermine human autonomy is paramount. At AutonomyAI, a company specializing in autonomous AI solutions, the team was focused on developing AI systems that empowered individuals rather than replacing them.

One approach was to incorporate ethics into the AI system design. The team developed a set of ethical guidelines that the AI system would follow when making decisions. Another approach was to promote reflective judgment in decision-making. The AI system would provide recommendations, but the human user would decide.

By incorporating ethics into the AI system design and promoting reflective judgment in decision-making, the AutonomyAI team developed AI systems that supported rather than undermined human autonomy.

The Role of Humans in the AI Decision-Making Process

Human oversight and intervention are crucial in the AI decision-making process. At HumanAI, a company specializing in human-AI collaboration, the team focuses on developing strategies for effective collaboration.

One strategy was to design the AI system to explain its decisions. This helped to build trust and confidence in the AI system and allowed human users to understand better the decisions being made. Another strategy was to implement a feedback loop that allowed human users to provide feedback on the AI system's decisions. This helped to improve the AI system's decision-making capabilities over time.

By focusing on effective human-AI collaboration, the team at HumanAI developed AI systems that enhanced rather than replaced human decision-making.

Impact of AI on Human Agency

The dependence on AI for decision-making can lead to human agency and autonomy loss. At AgencyAI, a company specializing in AI and human agency, the team was exploring the potential consequences of this dependence and ways to mitigate it.

One potential consequence was the erosion of human decision-making skills. If individuals rely too heavily on AI for decision-making, they might lose the ability to make decisions independently. Another potential consequence was the loss of human autonomy. If AI systems made all the decisions, individuals might lose control over their lives.

The AgencyAI team developed strategies to maintain and enhance human decision-making skills in an AI-dominated world to mitigate these potential consequences. These included providing training and education on decision-making, implementing features that encouraged reflective judgment, and promoting a balanced approach to AI and human decision-making.

AI and Decision-Making Skills

The reliance on AI for decision-making can impact the development of human decision-making skills. At SkillAI, a company specializing in AI and decision-making skills, the team was focused on developing strategies to maintain and enhance decision-making skills in an AI-dominated world.

One strategy was to provide training and education on decision-making. This included workshops and online courses that taught individuals how to make better decisions, even in the presence of AI systems. Another strategy was implementing features in the AI system that encouraged reflective judgment. This included explaining the AI system's decisions and allowing human users the final say.

The SkillAI team developed AI systems that supported rather than undermined human decision-making capabilities by focusing on maintaining and enhancing human decision-making skills.

AI, Autonomy, and Governance

In a world where artificial intelligence has become ubiquitous, governance and authoritarianism have profound implications. Governments and regulatory bodies are crucial in ensuring that AI supports rather than erodes human autonomy and decision-making. Imagine a country where the government has deployed AI systems to monitor and predict criminal activities.

While this may decrease crime rates, it also raises concerns about surveillance and individual freedom. There might also be concerns about the government misusing this technology to suppress political dissent or target minority groups.

Governments and regulatory bodies need to establish clear guidelines and regulations to ensure that AI is used ethically and responsibly and does not undermine human autonomy and decision-making. This involves creating frameworks for the ethical use of AI, ensuring transparency in AI algorithms, and establishing accountability mechanisms for AI systems.

Additionally, there should be a constant dialogue between technology developers, policymakers, and the public to address concerns and develop fair and equitable policies.

Legal and Regulatory Considerations

As AI systems become more sophisticated and integrated into our daily lives, legal and regulatory challenges associated with AI decision-making become more complex.

For example, consider a scenario involving an autonomous vehicle in an accident. Who is liable for the damages: the owner of the vehicle, the manufacturer, or the developer of the AI system? There might also be concerns about the data used by the AI system. Was it obtained ethically, and is it being used appropriately?

Accountability, liability, and data protection issues need to be addressed to ensure that AI systems are used responsibly and ethically. This involves creating legal frameworks that define the responsibilities of different stakeholders, establishing mechanisms for data protection, and developing standards for the ethical use of AI.

Moreover, as AI systems become more integrated into critical areas such as health care and transportation, there will be a need for more stringent regulations and standards to ensure safety and accountability.

Trust in AI Decision-Making

Building trust in AI systems is crucial for their widespread adoption and success. Imagine a doctor who is using an AI-powered diagnostic tool. If a doctor does not trust the tool's recommendations, she may not use it, leading to suboptimal patient outcomes. Similarly, if a judge does not trust an AI system used for sentencing recommendations, he may not use it, leading to potentially unfair sentences.

Transparency, explainability, and accountability are key factors in building trust in AI systems. Organizations need to ensure that their AI systems are transparent in their decision-making processes, can be explained to nonexperts, and are held accountable for their decisions. This involves developing explainable AI algorithms, creating mechanisms for accountability, and ensuring transparency in AI systems.

Additionally, there needs to be ongoing education and training for users of AI systems to understand how they work and how to interpret their recommendations.

Bias and Fairness

Ensuring fairness in AI decision-making is a significant challenge. AI systems are trained on data that may contain biases, which can lead to unfair decisions. For example, consider a hiring algorithm trained on historical data where certain groups of people were unfairly favored over others. If this bias is not addressed, the algorithm may continue to make unfair hiring decisions.

There might also be concerns about the impact of biased AI systems on marginalized communities. For example, an AI system used in criminal justice may disproportionately affect people of color if it is trained on biased data.

Organizations need to be aware of the risks of reinforcing existing biases in data and develop strategies for identifying and mitigating biases in AI systems. This involves conducting bias audits of AI systems, developing fair algorithms, and ensuring that AI systems are tested and validated for fairness.

Furthermore, it is essential to have a diverse group of people involved in developing and testing AI systems to ensure that different perspectives are considered and potential biases are identified and addressed.

Global Perspectives on AI and Autonomy

Different regions and cultures approach the issues of AI, autonomy, and decision-making differently. For example, in some countries, there may be a higher level of trust in AI systems and a willingness to delegate more decisions to machines.

In others, there may be a preference for human decision-making and a skepticism toward AI. Understanding these differences is crucial for global businesses and international cooperation. For example, a company that develops AI systems may need to tailor its products to the preferences and attitudes of different regions.

Organizations must be aware of the cultural and regional differences in attitudes toward AI and autonomy and tailor their approaches accordingly. This involves researching attitudes toward AI in different regions, understanding the cultural and regulatory differences, and developing strategies to address these differences. Additionally, there should be efforts to promote international collaboration and standards for the ethical use of AI and autonomy to ensure that these technologies are used responsibly and ethically worldwide.

Future Trends

As technology evolves, automated and autonomous systems are expected to become more sophisticated and capable. Imagine a future where autonomous vehicles are the norm, and AI-powered robots perform most manual labor.

This will have significant implications for businesses and cybersecurity. For example, there might be concerns about the security of autonomous vehicles and the potential for hacking. Organizations need to stay ahead of the trends and be prepared for challenges and opportunities.

This involves staying informed about the latest developments in AI and autonomy, developing strategies to leverage the benefits of these technologies, and addressing potential challenges. Moreover, organizations need to be proactive in shaping the future of AI and autonomy by investing in research and development, collaborating with other organizations and governments, and advocating for responsible and ethical use of these technologies.

The Next Phase in Evolution

Organizations must evolve to stay competitive as AI and autonomous systems become more integrated into our daily lives. This may involve rethinking organizational structures, processes, and the role of human contributions. For example, consider a company implementing AI-powered decision-making tools across its operations. This may lead to more efficient decision-making and operations. It may also require reevaluating the role of human employees and how they can best contribute to the organization's success.

There might also be concerns about the impact of AI and automation on employment. Organizations need to carefully consider the implications of AI and autonomy for their operations and develop strategies to leverage the benefits while mitigating the challenges. This involves developing strategies for human-AI collaboration, reevaluating organizational structures and processes, and addressing the potential impact on employment.

HUMAN-CENTRIC AI

Machines were seen as vast oceans of logic, tirelessly "crunching" numbers and managing data. They operated in binary rhythms: 1s and 0s, on and off, yes and no. Humans navigated life driven by emotions, desires, aspirations, and an innate need for connection. It was a time of two solitudes: the logical machine and the emotional human.

Yet, these two seemingly disparate entities started conversing, understanding, and even collaborating. Imagine a grand orchestra. Each musician brings a unique sound to the ensemble. The violinists make sweet melodies, the trumpeters make bold declarations, and the drummers make rhythmic beats. Without a conductor to guide them, to interpret the emotion behind each note and to harness the collective talent, the symphony remains just a cacophony of sounds.

In our narrative, humans are the conductor. We bring the nuanced understanding and the emotional depth. With its vast capabilities, AI serves as our orchestra, ready to amplify our intentions, understand our desires, and collaboratively create a masterpiece. It is not just about algorithms serving data, but about AI understanding and respecting what makes us human.

Human-centric AI is more than just a technological advancement; it is a philosophy. It states that for AI to benefit humanity, it must be designed with human well-being at its core. AI must not replace humans but amplify their capabilities, and not dictate their behavior but to assist with their goals. AI must listen, learn, and love, serving human needs and ensuring our collective well-being.

As we embark on this journey together, let us remember that AI and humans are not adversaries but partners in life, each enhancing the other's strengths. It is clear that the possibilities are boundless when human creativity meets AI capability.

Designing AI for Human Well-Being

Our humanity is still important, even as the world becomes increasingly dominated by machines and algorithms. As AI intricately weaves itself into our lives, we must ensure this technology enhances and elevates the human experience.

Imagine a future where the tools and gadgets we interact with daily are acutely aware of our emotional and physical needs. A world where AI does not just respond to commands but understands the subtle undertones of our voices, the unspoken words behind our texts, or problems we have after a long day.

Picture waking up in the morning, where your AI system is not just an alarm clock, but an entity that comprehends your sleep patterns. It knows when you have had a restless night, adjusting the ambiance of your room, playing soft melodies, or recommending a soothing meditation to ensure your day begins on a positive note.

Then, there are our social connections. In an age where digital interaction often surpasses physical, AI is pivotal in ensuring we do not lose the essence of genuine human connection. Beyond suggesting friends or groups based on interests, the AI of tomorrow dives deep into understanding our personalities, values, and dreams. It fosters meaningful connections, enriching our lives with interactions that resonate with our very core.

The true miracle unfolds in the domain of health. AI becomes a silent protector, no longer confined to just fitness trackers counting steps or calories. It observes, learns, and gently reminds us to take care of ourselves. Whether detecting early signs of mental stress, recommending slight changes in diet, or encouraging breaks to combat potential burnout, the AI stands as a sentinel, safeguarding our well-being.

As this mosaic takes shape, the truth is clear. Advancing AI is not about building formidable devices; it is about developing instruments that naturally integrate into our existence, recognizing our dreams, ambitions, and, most importantly, our yearning for authentic human connection. The evolution of AI is about partnership, not supremacy, where tech and humans progress together, helping each other.

Ensuring AI Serves Human Needs

As we navigate the digital renaissance, it becomes increasingly evident that the quintessence of AI is its potential to dovetail with and enhance the intricate weave of human needs and desires. With its rich cultures and backgrounds, our global society finds itself on the precipice of a technological epoch where the machines we have built could transcend their roles as tools and become companions in our shared journey of evolution. Imagine this: when you ask your phone about today's weather, you are not just talking to a machine. You are using something made to help you in your everyday life. It is more than just making tasks easier; it is about understanding what we need and want.

Our basic human needs, from the elemental, like food and shelter, to the more intricate, like companionship and purpose, have always guided our progress. In this landscape, AI emerges not as a just as technology, but as a resource to meet these needs with great precision and empathy. Take, for instance, the potential of AI in health care. Beyond robotic surgeries and diagnostics, AI could personalize health care to an individual's unique genetic and environmental makeup, predicting ailments before they manifest and offering solutions tailored to individual nuances.

Similarly, AI's promise in education is not confined to virtual tutors or digitized classrooms. Envision an educational ecosystem where learning is dynamic, adapting in real-time to a student's cognitive strengths and weaknesses. For example, a child in a remote part of Africa would have access to the same caliber of education as one in Silicon Valley, all thanks to the democratizing power of AI.

With these exhilarating possibilities comes responsibility. As we embed AI deeper into our society, the guiding principle should be a symbiotic relationship between humans and machines. It is essential to ensure that AI systems remain anchored in human values, ethics, and needs as they grow more autonomous. These systems must be built with consideration for the intricacies of human emotion, cultural context, and ethical considerations.

Moreover, the economic implications of AI should be keenly observed. While it offers unparalleled efficiencies and cost savings, a balanced approach is needed to ensure it does not exacerbate societal inequalities. The dividends of AI should be distributed in a manner that uplifts all strata of society.

In an era dominated by AI, our focus should remain unwavering: to craft systems that, above all, serve humanity. The litmus test for any AI advancement should be its tangible impact on human well-being and its ability to augment our shared human experience. Only by intertwining technology with these foundational principles can we truly realize the transformative potential of AI.

AI and Emotional Intelligence

We can feel oppressed by data and logic, and emotion remains an irreplaceable essence at the core of our existence. It is what defines our most intimate moments, drives our deepest convictions, and influences our most pivotal decisions. As we enter an era of unprecedented technological advancement, how do we ensure that our creations do not miss out on this quintessential aspect of being human?

Consider the human experience. We smile when we are happy, tear up during moments of sadness, and wear myriad expressions to convey anger, surprise, or contemplation. These gestures, subtle as they may be, are powerful communicators. Historically, machines operated in black and white, understanding commands, but remaining oblivious to the undertones of human emotion. Imagine a future where AI can understand these emotional cues and respond to them in kind.

Integrating AI with emotional intelligence is a necessity. Picture a virtual assistant that can detect the stress in a student's voice and offer encouragement. Imagine a health care robot that perceives a patient's anxiety and modifies its approach to be more comforting. These scenarios, once relegated to science fiction, are on the horizon thanks to advancements in emotion AI.

Harnessing technologies like facial recognition, voice pattern analysis, and natural language processing, researchers are crafting AI systems that can recognize and interpret human emotions. But the mission does not end at mere recognition. The true challenge lies in ensuring AI systems respond empathetically, forging genuine and meaningful connections. It is about striking the perfect balance: where AI does not overstep its bounds and become intrusive but does not remain a detached observer.

We must reflect on our responsibilities. Emotionally intelligent AI promises many benefits, from personalized customer experiences to enhancing mental health support. However, it also poses questions about privacy, ethical considerations, and the very nature of human-machine relationships.

As AI evolves and takes on aspects of emotional intelligence, it mirrors our own journey. Just as we grow, learn, and mature by understanding and interacting with our emotions, so too will AI. The future is not just about creating intelligent machines; it is about sculpting AI that understands and resonates with the heartbeat of humanity. In this endeavor, we are not just engineers and developers; we are artists, painting a canvas of a future where technology and human emotion walk hand in hand.

Collaborative AI

Collaboration has always been at the center of human progress. Our greatest accomplishments have historically come from pooling our collective intelligence, skills, and resources. As we stand at the dawn of a new technological age, a new partner is joining our collaborative efforts: artificial intelligence. How does one ensure that this collaboration is genuinely synergistic?

Imagine an architect skilled in designing aesthetically pleasing structures, but occasionally challenged by complex mathematical calculations. Now, pair this architect with an AI system that can instantly compute structural loads or simulate environmental effects on a building. The architect provides the creative vision; the AI delivers precise calculations and simulations. Together, they produce a design that is both stunning and structurally sound.

This is the essence of collaborative AI. It is not about replacing humans but improving their lives. From health care to business strategy, melding human intuition with AI precision can lead to outcomes neither could achieve alone.

In the medical field, doctors equipped with AI diagnostic tools can more accurately identify diseases, drawing from vast medical databases in real time while also considering the nuanced symptoms of the individual patient. Researchers can manage massive datasets with AI's assistance, uncovering patterns or insights that might have otherwise remained hidden.

Yet, the success of collaborative AI hinges on several factors. Systems must be intuitive, adapting to human workflows rather than forcing humans to adapt. Trust is paramount: professionals must understand and trust the AI's recommendations, ensuring it is a transparent and reliable partner.

Equally vital is these AI systems' ongoing training and fine-tuning, constantly learning from human expertise and feedback. This dynamic ensures the AI remains a valuable collaborator, always complementing human expertise without overshadowing it.

As we move forward, the potential of collaborative AI seems boundless. It requires careful design, mutual respect, and a clear understanding of roles. When it is done correctly, the human-AI partnership promises a future where combined strengths lead to unparalleled achievements.

Explainable AI (XAI)

Artificial intelligence affects many industries, transforming sectors and reshaping decision-making. How can we trust decisions made by machines if we do not understand how they arrived at them? The answer to this puzzle lies in the domain of explainable AI, often called XAI.

Imagine for a moment a seasoned physician presented with a diagnosis from a new AI tool. The tool claims that a patient, based on numerous variables and data points, has a certain medical condition. The doctor wonders, "Why? What did the AI see or interpret that led to this conclusion?" It is insufficient for the doctor (or the patient) to merely accept the AI's assertion; they need an explanation, a rationale.

That is where XAI becomes important, as bridges the confusing machine learning algorithms and the clarity humans need. Instead of black-box decisions where outcomes emerge without insight into the process, XAI shows the inner workings of these complex systems. It enables AI to articulate its decision-making process in terms humans can understand, leveraging visualizations, straightforward language, or comparative data points.

Why is this crucial? First, it is about trust. When users, whether doctors, engineers, or everyday consumers, understand the "why" behind an AI's decision, they are more likely to trust and adopt the technology. This trust is not just about comfort; it is about ensuring safety, especially in high-stakes areas like medicine or autonomous vehicles.

Furthermore, explainability can be a tool for refining AI itself. By understanding how a model reaches its conclusions, developers can identify biases, errors, or oversights in the algorithm, leading to more accurate and ethical AI solutions.

The journey toward widespread AI adoption is paved with challenges and questions. With the evolution of concepts like XAI, we are moving closer to a future where AI is not just powerful, efficient, transparent, and trustworthy.

Integrating these systems into society is not merely about what they can do; it is about ensuring they can communicate their actions in a way that aligns with our need for understanding and trust.

Personalized AI Experiences

With the ubiquity of smartphones, wearables, and smart home devices, we generate an astounding amount of data daily. The question is not just about handling this vast amount of information but how to transform it into something meaningful for each individual, which is the domain of personalized AI experiences.

Imagine browsing an online bookstore. Among the millions of titles available, which ones would interest you the most? Instead of sifting through endless lists, wouldn't it be delightful if the store could present a curated list as if a close friend had handpicked them just for you? This is the promise of personalized AI. Using intricate algorithms, AI assesses patterns, behaviors, preferences, and past choices, curating experiences tailored just for you.

This is not only a convenience: it is an enhancement of the quality of our digital interactions. Consider a student using an educational app. A one-size-fits-all approach might leave some struggling and others bored, but an AI that adapts and understands each student's learning pace, strengths, and weaknesses can offer a customized learning journey, maximizing potential and keeping engagement high.

It goes beyond mere commerce or education. Consider health apps that adapt workout regimes based on an individual's progress, media platforms that suggest content in tune with one's mood, or smart homes that adjust the ambiance based on the resident's comfort preferences.

However, this personalized world is not without its challenges. The balance between customization and privacy is delicate. Users want experiences that resonate with their personal tastes, but they also want their data to be protected. The app must balance between offering tailored services and respecting individuals' data rights.

The potential is undeniable. As AI continues to evolve and understand us better, the line between machine interaction and human behavior may become unclear. In the future, AI will not just be about algorithms and computations. It will be about delivering deeply personal experiences, as if each digital interaction was designed with just *you* in mind.

Accessibility and Inclusivity in AI Design

We must uphold the principle that the innovations of tomorrow must be for everyone, not just a select few. That means including accessibility and inclusivity in AI design.

Consider a bustling city. Different people navigate their daily routines within its streets and avenues. A young entrepreneur races to her next meeting, an elderly gentleman enjoys a leisurely walk, and a visually impaired student confidently uses his "smart" cane to guide him to university. Each of these individuals, with their unique life stories and challenges, is a testament to the diversity of human experience. AI, with all of its potential and power, should be a tool that acknowledges and serves this myriad of human experiences.

Regarding accessibility in AI, it is about ensuring that people with disabilities have AI tools that cater to their needs. Voice assistants, for instance, can become indispensable tools for visually impaired individuals. Meanwhile, gesture-based controls can empower those with hearing impairments to interact with their devices profoundly and meaningfully.

However, inclusivity in AI design can be defined more broadly. It is about cultural understanding, linguistic capabilities, and even socioeconomic considerations. An AI chatbot should be able to understand various accents and dialects, not just the mainstream ones. An AI-driven educational platform should cater to different learning styles and paces, recognizing that not everyone absorbs information the same way.

It is crucial to remember that the path to inclusive AI is not merely about software updates or tweaking algorithms. It starts with the humans behind the technology. Diverse teams, individuals of different backgrounds, abilities, and perspectives, can better envision and create AI tools that reflect the world's richness and diversity.

AI is not just about codes, algorithms, or cutting-edge technology. It is a reflection of our societal values and our collective aspirations. In a world of diversity, our AI solutions must incorporate inclusivity, ensuring no one is left behind in this technological renaissance.

Human Oversight and Control

The advancement of AI conjures up imagery from the annals of science fiction: autonomous robots, machines making decisions without human intervention, and sometimes, dystopian futures where machines reign supreme. No matter how advanced AI becomes, however, it should continuously operate under the vigilant eye of human oversight and control.

Imagine a bustling hospital where an AI system is responsible for monitoring patients' vitals amid the hum of machines and the urgency of medical procedures. While this system can detect abnormalities faster than humans, what happens when it recommends a treatment? Should it be followed blindly, or should a seasoned doctor always have the final say? The answer leans toward the latter. Medical treatments are a matter of life and death, so human judgment remains irreplaceable because it is based on years of experience and intuitive understanding.

Consider autonomous vehicles. As these cars operate on highways, using sensors and algorithms to navigate, there is comfort in knowing that a human can always intervene, take the wheel, and override the AI, especially in unpredictable situations that require split-second decisions infused with empathy, ethics, and understanding.

This principle of human oversight is not just about safety; it is deeply intertwined with trust. As AI systems become an integral part of industries ranging from health care to finance, users need assurance. They need to know that behind the complex calculations and data analyses, there is a layer of human judgment that ensures the machine's decision aligns with our broader societal values.

Moreover, with this oversight comes accountability. When AI-driven decisions go awry, as they inevitably will at times, it is essential to have a framework that allows for responsibility, redress, and rectification. Humans should have the control to intervene and the duty to answer for AI's actions.

The AI landscape is undoubtedly exhilarating and filled with potential and promise. Yet, as we march forward, we must anchor ourselves to the belief that these tools, no matter how sophisticated, remain just that — tools. It is up to us, the humans behind the machines, to ensure they serve us ethically, responsibly, and under our discerning control.

Continuous Learning and Feedback Loops

As we explore the relationship between humans and artificial intelligence, one aspect resonates profoundly: the cycle of continuous learning and feedback. The beauty of AI is not just in its ability to compute at phenomenal speeds or analyze vast data sets with ease. Its true power lies in its capability to evolve, refine itself based on human feedback, and grow aligned with our values and aspirations.

Think back to the early days of any technology, from the first cars to the initial versions of your favorite apps. They were often lacking features and were not very user-friendly. They improved with each iteration based on feedback and real-world usage. Now, magnify that potential for growth by several orders of magnitude, and you have got AI.

Unlike traditional software, AI systems have the inherent ability to fine-tune their operations based on feedback. Consider a voice assistant, for instance. The first time you ask it to play your favorite song or set a reminder, it might falter, misunderstanding your accent or phrasing. Each interaction and each correction become a learning opportunity. Over time, through repeated feedback, it begins to understand you better, making fewer mistakes and providing a smoother experience.

This feedback cycle and continuous learning ability hold profound implications. The stakes are significantly higher in sectors like health care, where AI assists with diagnoses or suggests treatments. Here, feedback loops are not just about refining an algorithm, but could mean the difference between accurate and potentially erroneous medical advice. Human experts need to consistently evaluate, correct, and train these systems, ensuring they do not just repeat past mistakes but learn and adapt from them.

This dynamic interplay also reinforces the partnership ethos between humans and machines. AI is not a static entity to be deployed and forgotten. It is a vibrant, ever-evolving tool that relies on human interaction for its growth. As these systems learn from us, we too learn from them, understanding their strengths, recognizing their weaknesses, and working together to harness the best of both worlds.

At the heart of this lies a promise: a commitment to making AI more brilliant in a computational sense and wiser and more attuned to human values and needs. With feedback loops and continuous learning, the journey of AI is not a sprint but a marathon, where each step, each iteration, brings us closer to a harmonious blend of technology and humanity.

SOCIETAL IMPACT OF AI

Artificial intelligence is a technological advancement designed to emulate human cognition, enabling machines to perform tasks that typically require human intelligence. This includes problem-solving, recognizing patterns, understanding language, and making decisions. As AI's capabilities have expanded, so too has its influence on various aspects of society. This influence, both positive and negative, is what we refer to as the societal impact of AI.

Visualize an academic institution where AI systems are entrusted with tailoring educational experiences. While such a system can optimize learning paths more efficiently than its human counterpart, how much influence should the system have on determining educational pathways? Should an AI, with its vast data, dictate a student's career or should an experienced educator who understands the nuances of human ambition and potential be the anchor? Human intuition seems to be the better choice in this example.

Let us consider the influence of AI in our cities. AI-driven systems manage traffic, predict criminal activity, or even influence policy-making. Should humans guide AI systems in situations with possible moral, cultural, and ethical dilemmas? Decisions that shape lives and cultures require human intervention, with its lived experience and collective wisdom.

This "human touch" in AI is not just for sentimentality; it is important for trust. As AI influences sectors from urban planning to arts, the people who must abide by its effects need reassurance about those decisions. This reassurance includes human judgment, a safeguard ensuring an AI system's decisions align with our societal mores.

Accountability is an important aspect of any influential system. When AI-driven outcomes falter or misalign, there should be a framework for intervention, accountability, recalibration, and ethical realignment. Humans should not merely be reacting to safety issues but actively involved in guiding AI decisions.

The potential of AI is undeniably vast. These possible future innovations, regardless of their brilliance, are "instruments" in our societal "orchestra." Humanity must ensure this "symphony" resonates with harmony, ethics, and shared vision.

Impact on Employment and Economic Inequality

Progress, driven by artificial intelligence, is reshaping employment and economic structures. You can likely imagine a factory floor where robotic arms, guided by AI, assemble products with precision and stamina. This activity reveals that AI deployment is not just about efficiency and productivity; it is about the changing job roles and economic implications that affect other people.

The effect of AI on jobs shows how complex the issue is. There is an undeniable boon of AI: creating new job domains, fostering innovation, and opening new areas in technology and science. As with any profound change, however, there are some challenges. Traditional jobs that have been the backbone of industries for decades face transformation or even obsolescence. The equilibrium between man and machine is delicate. While AI ushers in an era of unparalleled possibilities, it poses questions about economic disparities and workforce displacement.

Adaptability and lifelong learning are important parts of modern work. As routine tasks are delegated to AI, the human workforce must enter roles that demand creativity, empathy, and strategic thinking, skills that AI cannot replicate (at least, not yet). The AI that disrupts jobs also holds the potential to educate and re-skill the workforce, tailoring learning to individual needs and market demands.

Society must make some important decisions about using AI. People can use it to increase economic inequality, widening the chasm between those who can harness AI and those who cannot. People can also choose to put forth a collective effort to ensure that the benefits of AI are accessible and equitable, that the workforce is prepared for the future, and that economic growth does not become unfair.

Effects on Social Interactions and Cultural Values

As AI becomes part of society, it creates a complex "story" of interaction and culture. Imagine a world where AI-driven social platforms connect us and influence the nature of our interactions. Algorithms curate our newsfeeds, shape our opinions, and even mediate our relationships. Here, AI is not just a tool; it becomes a mediator of human connection.

This issue is not just about technology's influence on social media sites; it is about how it is changing human relationships and cultural values. AI breaks down geographical barriers in virtual classrooms and online forums, allowing for the sharing of ideas and perspectives. Yet, AI's algorithms can create echo chambers, reinforcing our biases and narrowing our worldviews.

Diversity, empathy, and authenticity are important cultural values. As AI becomes ubiquitous in social media, entertainment, and daily communication, it subtly challenges these values. AI can foster an environment like a global "village," promoting understanding and inclusiveness, or it can damage societies, amplifying divisions and creating a polarized world.

The responsibility for the impact of AI is on creators, users, and regulators of AI. The challenge is to harness AI to enrich social interactions without hurting human empathy and cultural diversity. It is about finding balance: making AI support and enhance human interaction rather than replacing or dictating it.

The future of AI depends on today's choices: how we design AI, the ethical frameworks we build around it, and the cultural values we embed within it. AI can either be a catalyst for a more connected, understanding, and empathetic society or a divider that deepens societal rifts.

ETHICAL CONSIDERATIONS IN AI DEVELOPMENT

Think of an AI developer's workspace not as a cluster of codes and algorithms, but as a place where ethical dilemmas and technological prowess intertwine. Every line of code, every algorithmic decision, carries certain ethical implications. The developers are not just engineers of technology; they are "architects" of ethical frameworks. Their decisions have an effect on society, impacting everything from individual privacy to societal norms.

The principle of "do no harm" is critical to the development of AI systems. With its profound ability to learn, predict, and act, AI holds immense power. With great power comes great responsibility. How can AI be prevented from unintentionally perpetuating biases or infringing individual rights? There are important considerations in regard to the difference between enhancing human capabilities and infringing upon human autonomy.

Consider an AI system designed for predictive policing. While it aims to reduce crime, it also risks reinforcing existing societal biases. Now consider an AI health care system that proposes treatment plans; it must do so without impinging on patient privacy or autonomy. Here, the ethical conundrum lies in ensuring that AI acts as an aide to human expertise, not as an authority.

Transparency forms another pivotal consideration. As AI systems become more complex, understanding their decision-making processes becomes crucial. Ethical AI development requires that these systems not be "black boxes." Instead, they should be interpretable and explainable, enabling users to understand and trust their actions and decisions.

Accountability is another important consideration. When AI errs, who is held responsible? There must be clear determinations of accountability. Accountability is not just about fixing blame; it is about creating systems that can learn from their mistakes, where developers, users, and regulators are responsible for guiding AI toward beneficial outcomes for society.

Developers of AI systems must also explore the balance between innovation and regulation. Ethical AI development is not about stifling innovation with overly stringent rules. Instead, it is about guiding innovation that respects human dignity, rights, and societal values.

The ethical considerations in AI development are part of an ongoing discussion. It involves navigating the complex interplay between technological advancements and human values. As AI continues to evolve, so must our approach to its ethical implications, ensuring that it is used for beneficial purposes, enhancing our lives while upholding our values.

Ethical Responsibilities of AI Developers and Operators

Ethical responsibilities are like guides for developers and operators. These responsibilities involve adhering to the technical specifications and upholding moral principles.

Pretend you are an AI developer. Each algorithm you create is not merely a set of instructions; it is a decision that could impact lives. You are not just coding but embedding ethical judgments into your program. For instance, designing an AI for loan approval is not just about assessing creditworthiness. It is about ensuring the AI does not perpetuate societal biases, unfairly disadvantaging certain groups. The responsibility extends to understanding the societal contexts and nuances, ensuring the AI's judgment is fair and just.

Now let's consider the role of an AI operator, where the responsibility shifts to how the system will be applied. The operator must ensure the AI systems are used ethically, respecting user consent and privacy. The AI operator must leverage the AI for business benefits and respect individual rights.

Let us consider an AI used for personalized advertising. The operator's ethical duty involves ensuring that user data privacy is not compromised while targeting ads and that the AI does not become intrusive or manipulative. The operator's responsibilities encompass a broad spectrum of duties: from actively seeking to mitigate biases in AI systems to ensuring transparency and explainability in AI decisions, respecting privacy and data protection laws, and being vigilant about AI's impact on human jobs and societal norms.

Ethical Guidelines and Frameworks for AI Development

Ethical guidelines and frameworks shape AI's development. These guidelines form a philosophical foundation that ensures AI's power is harnessed for the greater good.

Let us consider a different scenario: a council of AI ethicists and technologists are drafting a framework. It is a scene where philosophical theories meet algorithmic complexities. These people must focus on fairness and non-discrimination. The framework they design stipulates clear criteria to prevent AI from becoming biased, ensuring algorithms treat all users equitably, irrespective of race, gender, or socioeconomic status.

This framework must include transparency and accountability. AI systems should be transparent enough for users to understand how decisions are made. If an AI denies a loan or diagnoses a disease, users should be able to see the "why" behind these decisions. Accountability is equally crucial. The guidelines ensure that there is always a human accountable for the AI's decisions, someone who can answer for its actions and rectify mistakes if needed.

Privacy protection forms another part of the framework. There must be strict protocols for data usage, ensuring that AI respects user consent and data privacy and does not become a tool for unwarranted surveillance.

Moreover, the framework must incorporate safety and security, setting standards to ensure AI systems are robust against hacking and misuse and function reliably without causing unintended harm.

Finally, there must be a focus on societal and environmental well-being. This part of the framework ensures that AI development is aligned with sustainable practices and that its deployment considers societal impacts, like job displacement and social disruption.

These ethical guidelines and frameworks guide the AI through challenging moral dilemmas and ethical quandaries, ensuring it can provide a useful service and innovation without becoming unethical.

THE ROLE OF REGULATION IN AI

Consider the following analogy: AI is not a solitary musician but a full-fledged orchestra. Each instrument in this analogy (such as a health care algorithm diagnosing cancers, a financial model predicting market trends, and an autonomous car weaving through city streets) holds immense potential. The benefit of each lies not in the individual notes made by the "instruments," but in their harmonious interplay. If each AI system is an instrument, then who is the "conductor" ensuring the orchestra plays together? Here, the "conductor" would be *regulation*.

Regulation does not stifle innovation; it sets the boundaries within which the progress can flourish. As AI gains autonomy, its decision-making processes become more important. Regulation ensures these decisions remain transparent, fair, and accountable, like ensuring every instrument plays its part in an orchestra without overshadowing the others. Regulators set the rules so that developers, users, and everyone impacted by AI know their roles, as well as their rights and responsibilities.

Regulation is not merely a "conductor:" it also guards against unintended consequences. Biases in algorithms can create unfairness and discrimination. Regulation acts as a filter, ensuring AI does not injure fundamental human rights and privacy that are the foundation of our society.

The beauty of an AI system is its adaptability. Regulations must keep up with AI's ever-evolving progression. This means there must be a constant dialogue between technologists and engineers brainstorming new systems, lawmakers crafting nuanced regulations, ethicists refining the underlying guidelines, and the public giving feedback as attentive critics. It is an ongoing conversation, a collaboration to address new challenges and opportunities, similar to composing a piece of music that is both cutting edge and timeless.

Ultimately, regulation is not about stifling AI but ensuring that it aligns with human well-being and ethical principles. It is about creating a future where AI and humanity work together, each enhancing the other, which drives innovation developed with the values that make us human.

The Need for Regulatory Frameworks

Let us try another approach to think about AI and regulation. Consider a sprawling metropolis made not of steel and glass buildings, but data and algorithms: an "AI city." Like any dynamic city, this "AI metropolis" requires a comprehensive infrastructure and regulatory frameworks serve as the guidelines. These frameworks are not hindering the way the AI city operates, but helping to maintain a balanced urban ecosystem where innovation and ethical responsibility can exist together.

Think of the regulations as the laws guiding the AI systems in our imaginary AI metropolis. For autonomous vehicles, they help balance efficiency with safety. Regulations ensure the AI systems operate successfully in sensitive areas like health care, finance, and criminal justice in a fair, equitable, and non-discriminatory manner. Biases, which can be thought of as "roadblocks" in this AI city example, are assessed and mitigated to prevent any algorithm from becoming a source of societal friction.

AI frameworks must evolve with the technology. Regulations should not remain static but be dynamic, responding to AI's requirements. As machine learning algorithms become more autonomous and powerful, regulations must adapt to ensure they remain under the control of human values and societal norms. This relationship between the regulations and the AI systems should be seen as a constant dialogue between technological advancement and public interest, a negotiation between progress and preserving our core principles.

Ensuring Responsible and Ethical Development

Responsible AI development is a technical challenge that affects entire communities. AI developers and companies are tasked with constructing systems that are not just functionally advanced, but also morally sound and socially responsible. Every component must be infused with ethical considerations, from data collection to algorithm design. Fairness, transparency, and respect for privacy are important aspects of development.

Bias is a serious flaw in the development process, and so the AI system must be tested and any bias eliminated before full implementation can occur. Regular audits of AI systems can ensure they adhere to their ethical guidelines. Communication is critical to the community involved with the AI system, allowing users to understand how decisions are made and how data is used.

Successful deployment of the AI system depends on ensuring the AI complements human roles, not replaces them. AI should streamline tedious tasks and empower human potential, not usurp it. AI can handle menial tasks in workplaces, freeing up human minds for innovation and problem-solving. This synergy should be the guiding principle of AI system development.

However, after the AI system is implemented, constant monitoring is still required, ensuring the AI adheres to its ethical guidelines. Feedback mechanisms allow for human intervention when necessary. Ultimately, a human developer, user, or regulator must always be ready to make adjustments to the system and keep it aligned with human values.

The goal is not simply to harness AI's potential; it is to ensure that AI systems are used for the good of humanity. We seek innovation and AI systems that reflect our values of fairness, responsibility, and cooperation. AI regulation and ethics should help developers craft systems where the technology serves humanity.

2

AI APPLICATIONS AND REAL-WORLD CASE STUDIES

INTRODUCTION

In this chapter, we examine AI's diverse applications and real-world case studies. We will consider how AI transforms every sector it is used in and learn how it can affect innovation.

AI is versatile and can help motivate the discovery of breakthroughs, solutions, and new possibilities. AI's precision in uses like health care diagnostics and treatments and the ability to customize the AI makes it a useful tool to adapt, learn, and revolutionize various industries.

Let's consider the possible future impact of AI. For example, AI could result in a world where diseases are diagnosed with an accuracy that outpaces human capabilities, medical treatments are tailored to the individual's genetic makeup, and pandemics are better managed. In schools, we can imagine classrooms where learning is not a one-size-fits-all model, but a personalized journey tailored to each student's pace and style.

In finance, we might imagine an AI system that can detect fraud and manage risks with an efficiency unattainable by human standards. It may also be utilized as an intelligent adviser in trading, predicting market trends with algorithms more intricate than the human mind can understand.

Manufacturing facilities may be transformed by AI in terms of automation and process optimization. Thanks to AI's predictive capabilities, the manufacturing sector could experience a revolution in maintenance strategies, changing how downtimes are managed and uptimes could become longer.

In entertainment, AI will influence content creation, gaming experiences, and virtual reality. Its role in film production could result in unique scripts and visual masterpieces, as well as a redefinition of the post-production processes.

AI's influence may extend into space science, where it could aid in satellite data analysis, space exploration, and planning interstellar missions. AI's potential may create new opportunities off Earth.

AI's role in emerging fields like 3D printing, augmented reality, and cybersecurity reveals where its applications may be best utilized. AI can be used in the development of *digital twins*, which are computerized replicas of actual physical entities. The application of AI could revolutionize industries like manufacturing and urban planning.

This chapter contains a collection of case studies and shows the possibilities of AI in the future. AI has the ability to be a transformative force, reshaping industries, redefining paradigms, and enhancing technological evolution.

AI IN HEALTH CARE

Artificial intelligence (AI) in health care is a technological upgrade that is efficient; it is also redefining how we provide medical care. AI can augment human capabilities and change current reactive health care models to a predictive model of patient wellness. It can assist with treating illness as well as maintaining wellness, and make health care more personalized, accessible, and proactive.

Let us consider a future health care system that utilizes AI. AI systems can become an integral part of the decision-making process, empowering doctors with new insights from a large amount of medical data. In our future system, the AI algorithms can predict health issues before they arise, offer tailor made treatment plans based on a patient's unique genetic makeup, and provide real-time support to surgeons during complex procedures.

The notion of "one size fits all" will become obsolete in health care. By using big data and advanced analytics, AI can create personalized medicine treatments and provide medical advice uniquely tailored to the individual. This AI approach will result in better health outcomes and improve patients' quality of life.

Artificial intelligence can also be used in *wearables*, or devices that are attached to a person's body. Wearables can monitor a patient's vitals, and smartphone apps can help patients manage chronic diseases and provide telemedicine consultations. This technology makes health care more accessible to people, allowing for continuous monitoring and care, irrespective of geographical barriers.

The ethics of AI involved in health care should be carefully considered. How do we ensure fairness in AI diagnoses? What are the privacy implications of AI in health care? We must create a robust ethical framework for AI that respects patient autonomy and privacy while ensuring equitable access to these advanced health care solutions.

Health care professionals should prepare for a workplace that involves AI. We must rethink how medical education and training can include AI literacy, ensuring the health care workforce can work with and effectively utilize AI.

Integrating AI in health care will considerably change how the health care industry operates. It can create a health care system that is more efficient, accurate, humane, and personalized than the current system. AI in health care can provide patients with hope, drive innovation in medicine, and help make a healthier world for everyone.

AI-Driven Diagnostics and Treatment

AI algorithms, trained on vast datasets of medical images and records, can now identify diseases and conditions with a precision that rivals and sometimes surpasses human experts. From detecting early signs of cancer to identifying subtle changes in brain scans indicative of neurological disorders, AI enables earlier, more accurate diagnoses.

Take, for instance, the field of oncology. AI systems are being developed to analyze mammograms for signs of breast cancer with a degree of detail that escapes the human eye. These systems can recognize patterns indicative of tumor growth, potentially catching malignancies at stages early enough for treatment to be more effective and less invasive.

AI systems can analyze a patient's medical history, genetic information, and current health status to recommend personalized treatment plans. This approach considers the effectiveness of different treatments and factors in potential side effects, ensuring a tailored approach that maximizes efficacy while minimizing risks.

Predictive analytics is an area of analytics where AI analyzes trends and patterns in large datasets and makes predictions based on the likelihood of disease progression and response to various treatments. This approach helps create more effective treatment plans and aids in preventive care, identifying at-risk individuals, and facilitating early interventions.

While the potential of AI in diagnostics and treatment is immense, it is not without challenges. Issues like data privacy, the need for large, diverse datasets to avoid biases, and the integration of AI into existing health care systems are critical areas that need addressing. Moreover, the ethical implications of AI decision-making in health care require careful consideration and robust regulatory frameworks.

To fully realize the benefits of AI in diagnostics and treatment, there is a need for collaboration between technologists, health care professionals, and policymakers. Training health care workers to work alongside AI, developing AI guidelines, and investing in research are crucial steps toward an AI-enhanced health care future.

Personalized Medicine

AI systems have the unique ability to analyze vast amounts of data rapidly and accurately. This capability is particularly important in diagnostics, where AI algorithms sort through complex medical imagery, from MRI scans to X-rays, identifying patterns and anomalies that might escape a human. These systems, trained on extensive datasets, are becoming increasingly adept at identifying early signs of diseases such as cancer, often long before they become symptomatic.

AI's role in health care extends beyond diagnostics and treatment. It can reshape the approach to patient care. Let's consider the example of oncology, where AI systems analyze a patient's unique genetic information to recommend personalized treatment plans. This method, known as *precision medicine*, tailors therapies based on an individual's genetic makeup, lifestyle, and environment, which is very different from the current "one-size-fits-all" approach of traditional treatments.

Another AI advancement is in drug discovery and development. AI algorithms are accelerating this traditionally long and costly process by predicting how different drugs will interact with various targets in the body. This capability not only speeds up the development of new drugs but also enhances the safety profile of these medications by anticipating potential side effects and interactions.

With these advancements come significant ethical considerations. The reliance on vast datasets for training AI systems raises privacy concerns. How is patient data being protected? Moreover, the decision-making process of AI algorithms in health care is often a black box, lacking transparency. This obscurity poses a challenge in understanding and trusting the decisions made by AI, which is crucial in a field where human lives are at stake.

Despite these challenges, the potential of AI in health care to save lives and improve patient outcomes is immense. As technology continues to evolve, it may enhance medical care, making it more accurate, efficient, and personalized than ever before. For patients and health care providers alike, the future of AI-driven diagnostics and treatment may reshape health care as we know it.

Case Study: AI in Pandemic Management

The year 2020 was important in global health history because of the outbreak of COVID-19. As the world dealt with an unprecedented health crisis, it became evident that traditional disease surveillance and management methods were inadequate to address such a rapidly evolving pandemic. Artificial intelligence was successfully utilized in managing and mitigating the effects of the pandemic.

One of the most striking examples of AI's impact was seen in early detection and forecasting. As the virus began to spread globally, AI systems, trained on many data sources, including social media, travel patterns, and infection rates, became instrumental in predicting outbreak hotspots. These predictions enabled governments and health organizations to allocate resources more effectively and enforce targeted lockdowns and social distancing measures.

AI-driven analytics also played a critical role in understanding the virus's behavior. By analyzing thousands of scientific papers and datasets at an unprecedented speed, AI helped researchers identify potential treatment strategies and understand the efficacy of various drugs and therapies. This rapid analysis was crucial in developing treatment protocols in a scenario where time mattered.

Perhaps the most revolutionary use of AI during the pandemic was in vaccine development. Traditional vaccine development, a process that typically takes years, was expedited significantly. AI algorithms analyzed large databases of genetic data to identify potential vaccine targets. This process, which would have taken months or even years for people, was completed in days, allowing for the rapid development of effective vaccines.

Another vital area where AI proved useful was in health care delivery. With hospitals overwhelmed and health care workers under immense pressure, AI-powered tools and applications provided much-needed support. Chatbots and virtual health assistants offered preliminary diagnosis and advice, reducing the strain on health care facilities. AI also enhanced remote monitoring of patients, ensuring continuous care while minimizing the risk of virus transmission.

The use of AI in pandemic management was not without its challenges. Data privacy and security issues, as well as concerns about the ethical implications of AI-driven decisions, showed how underdeveloped the current ethical guidelines were. The accuracy of AI predictions also depended heavily on the quality and quantity of the data fed into these systems, highlighting the need for robust and diverse datasets.

The role of AI in managing the COVID-19 pandemic was a testament to its potential in facing global health crises. While it presented challenges, the benefits were undeniable. AI provided crucial insights and predictions and accelerated vaccine development, transforming the landscape of pandemic response.

This case study is a powerful example of how AI, when used responsibly, can be a powerful tool for safeguarding human health and well-being. The lessons learned from this experience will undoubtedly shape the role of AI in managing future health crises and may help us create better and more resilient responses.

The Future of Health Care

Let us consider another example: Imagine a world where a patient's treatment is tailored to uniquely address their individual genetic makeup, lifestyle, and environment. AI will normalize the application of precision medicine because of its ability to analyze large genetic datasets. Patients will receive medications and treatment plans specifically designed for their genetic composition, drastically improving treatment efficacy and reducing the side effects of treatments.

Diagnostic procedures will also be changed by AI. With their ability to discern patterns in large amounts of data that people may not otherwise notice, AI algorithms will be able to analyze medical images with unprecedented precision. From detecting the early onset of diseases like cancer to predicting potential future ailments, these algorithms will enable earlier and more accurate diagnoses, leading to better patient outcomes.

AI-powered virtual health assistants will always be available, providing constant health monitoring and personalized advice. These assistants will become personal health coaches, reminding patients to take medication, offering nutritional advice, and even providing mental health support. This constant monitoring and interaction will be particularly transformative for chronic disease management, offering patients a new level of independence and quality of life.

Surgical procedures may become more precise and less invasive with the integration of AI and robotics. Robotic systems, guided by AI algorithms, will perform complex surgeries with a precision that surpasses human capabilities. These surgeries will be more successful and significantly reduce recovery times and the risk of infection.

AI will also democratize access to health care. AI-powered telemedicine will provide high-quality health care to remote and underserved regions. AI tools will aid non-specialist doctors in remote areas in diagnosing and treating complex conditions, ensuring that all patients have access to excellent health care services.

Perhaps the most groundbreaking aspect of AI in future health care is its predictive capabilities. By analyzing patterns in health data, AI will predict illnesses before patients manifest symptoms, enabling preventive health care on a scale never seen before. This shift from reactive to predictive health care will save lives and significantly reduce health care costs.

The future of health care with AI could mean that health care is more accurate, accessible, personalized, and preventive. The use of AI to improve health outcomes may redefine what it means to be healthy and cared for.

AI IN EDUCATION

Artificial intelligence will have a strong effect on education. Let us consider another possible scenario. Envision a classroom where students all have access to a virtual assistant named "Ada." Each student is following a specially tailored lesson. The students not only obtain personalized learning for their unique abilities but are encouraged by the AI assistant. The human teacher in the classroom has a different role than that of a traditional instructor. He serves as a facilitator, assisting students with their personalized lessons. This scenario is not entirely fictional: it is already occurring in some classrooms.

Integrating AI into education has naturally started a debate between those who are optimistic about it and those who are skeptical. On one side, skeptics envision a classroom that is impersonal and subject to machine-based teaching. They voice concerns about the lack of a good learning environment with other people, especially the loss of the traditional teacher-student relationship.

AI's proponents see AI in the classroom as an "educational renaissance." They think of AI as a tutor that helps teachers with the mundane tasks of grading and planning. They believe the AI will free the teacher to focus more on nurturing, mentoring, and truly understanding each student. Proponents say that AI will give teachers more time to create individual learning journeys, crafting experiences that educate and inspire each student.

This debate is an important one. AI is not a replacement for human teachers. It is, rather, a complementary resource to assist both educators and students. AI supports learning inside the classroom. It can develop individualized learning paths while teachers guide, mentor, and support the emotional and intellectual development of the students.

The future of education with AI is a blend of human ingenuity and technological innovation. AI-assisted learning is personalized but also enhances the teacher's role in the classroom. As we consider the role of AI in education, we must consider how to best utilize its abilities with the human elements of teaching to create a more inclusive, engaging, and effective educational experience.

Personalized Learning Experiences

In envisioning the future of education, we see a dynamic transformation where AI does not just supplement but revolutionizes the learning experience. Imagine entering a classroom where traditional teaching paradigms are redefined, and every student embarks on a personalized educational journey tailored to their needs, learning styles, and passions.

In this AI-enhanced classroom, the "one-size-fits-all" model is obsolete. Instead, students are no longer confined to conventional learning methods. For example, some students struggle with understanding math. An AI assistant may be able to effectively communicate with some students by showing them 3D models or other types of examples that are more meaningful. Students benefit from the adaptation of the topic to suit their best learning style.

Auditory learners may particularly benefit from AI-developed simulations and dramatizations. In these types of lessons, the AI may create material where historical figures speak in their own voices and scientific experiments have life-like sound effects. Each lesson can make important facts easier to understand and improve student comprehension and desire to learn more.

Personalization in this AI-driven educational ecosystem extends beyond learning styles. It utilizes individual interests to nurture learning. For example, an AI assistant may develop lessons based on a student's interest in marine biology: the lessons are then delivered through an immersive VR experience of the Great Barrier Reef. Students can benefit from this tailored, immersive experience and participate virtually in global collaborative projects to obtain a broader understanding of the world.

AI-driven educational experiences are a convergence of technology and human-centric teaching. Students will utilize devices and VR headsets instead of textbooks and blackboards; instead of field trips and dissections, students can virtually visit anywhere and participate in virtual dissections. Teachers are the facilitators of these learning experiences.

AI-assisted learning can foster an environment where student curiosity is encouraged and creativity flourishes, and every student is empowered to chart their own unique educational journey. The individual subjects can be taught together to provide a more realistic experience of how to use the knowledge. This is the future of education with AI, of personalized learning experiences where each student's journey is as unique as they are.

AI in Educational Administration

Integrating AI into educational administration marks a significant shift in how educational institutions manage and streamline their operations. This transition is partly about improving efficiency, as well as reshaping the administrative landscape to support better educators, students, and the broader learning environment.

AI is a useful tool in educational administration, offering solutions to long-standing administrative challenges. Consider the immense amount of data educational institutions handle, from student records and academic progress reports to resource allocation and facility management. AI can help manage this data so administrators can make informed decisions, optimize processes, and predict future needs.

One significant application of AI in educational administration is student admissions and enrollment. AI systems can analyze application materials, predict enrollment trends, and even assist in identifying candidates who best fit the institution's ethos and academic standards. This speeds up the admissions process and adds a layer of analytical depth that human administrators alone might not achieve.

Another area where AI makes a substantial impact is in resource allocation. Through predictive analytics, AI can help institutions forecast future resource requirements, whether faculty hiring, classroom space, or learning materials. This proactive approach ensures institutions are ready for future challenges and changes.

AI can provide assistance for student support and services. From chatbots that provide 24/7 assistance to students for their queries to AI-driven platforms that track and support student well-being and academic progress, AI brings a more personalized and responsive approach to student services. These systems can identify students who might be academically or personally struggling and prompt timely intervention.

Financial management and budgeting in educational institutions are other critical areas where AI can be useful. AI can analyze spending patterns, predict future financial needs, and ensure that budgets are aligned with institutional goals and educational standards. This leads to more responsible and effective use of funds, directly impacting the quality of education provided.

AI can also play a role in maintaining and optimizing infrastructure, from energy management to facility maintenance. AI creates a sustainable and cost-effective learning environment by predicting maintenance needs and optimizing energy use.

AI's role in educational administration can assist the school with improved efficiency, foresight, and personalized service. It can help the administration become more responsive to the evolving needs of students and educators. This integration of AI supports a more dynamic, well-managed, and forward-thinking educational environment.

Case Study: AI in Online Learning Platforms

When the COVID-19 pandemic began, it caused a significant change in education. Traditional classrooms were closed, and many students utilized digital platforms. This change showed the problems with the current educational system and revealed latent opportunities. During this time, EduFuture, an emergent online learning platform, was useful in assisting the new educational experiences through the astute application of AI.

- *Strategic AI Integration:* EduFuture's platform employed a multifaceted AI strategy, each component serving a distinct role in enhancing the learning experience.
- *Data-Driven Learning Analytics:* EduFuture's AI suite relied on a sophisticated analytics engine that examined student engagement patterns and performance metrics. This engine tracked grades, and discovered how students interacted with course material, adapted to challenges, and emotionally responded while learning.
- *Dynamic Learning Pathways:* EduFuture's AI developed personalized learning journeys. These adaptive pathways ensured that each student encountered content at a pace and difficulty tailored to their unique learning abilities, and mirrored the attentiveness of a one-on-one tutor.
- *Interactivity and Support:* EduFuture introduced AI-driven interactivity. Students received instant, on-demand academic support through intelligent chatbots and virtual assistants, which allowed them to engage in a dynamic dialogue about learning.
- *Instantaneous Assessment:* EduFuture's AI algorithms provided real-time grading and feedback, akin to a coach providing instant playback analysis.

This immediate assessment allowed students to understand their performance gaps and successes.

- *Tangible Outcomes and Accessibility*: Students reported an increase in engagement levels, and this increased interest was attributed to the tailored learning experiences and interactive elements. EduFuture turned learning into an active, participatory process.
- *Elevated Academic Performance*: With personalized learning lessons, students experienced marked improvements in comprehension and retention. The AI-driven model catered to diverse learning needs, ensuring every student had the appropriate lessons.
- *Bridging the Digital Divide*: EduFuture expanded access to quality education through strategic partnerships, providing resources to those previously constrained by geographical or economic barriers.
- *Empowering Educators:* AI tools augmented the educators' roles, allowing them to utilize meaningful, personalized mentoring.

EduFuture's experience during the pandemic shows how AI can improve the educational landscape. When traditional learning models were no longer possible to utilize, EduFuture was able to deliver an AI learning experience that was inclusive, effective, and engaging. EduFuture's experience is an example of innovation and evidence of AI's potential to revolutionize education.

The Future of AI in Education

An AI-driven educational system could greatly enhance how educational institutions operate, teachers convey information, and students learn. AI can make learning more personalized, accessible, and efficient.

- *Revolutionizing personalized learning*: One of the most significant impacts of AI in education is the ability to provide personalized learning experiences. AI systems can analyze a wealth of data, including a student's learning style, strengths, weaknesses, and even emotional responses, to tailor educational content that resonates with each learner. Students who may struggle with traditional methods can learn through interactive 3D models that make abstract concepts tangible. For auditory learners, lessons can be made into an immersive experience.
- *Empowering educators and enhancing administration*: The role of educators evolves in this AI-enhanced future. Freed from administrative burdens like grading and attendance tracking, teachers can focus on mentoring and inspiring students. They become facilitators of a richer educational experience, leveraging AI-driven insights to meet individual student needs. For instance, a principal may utilize AI for data analysis, optimizing school operations and identifying student needs, thus transforming the school into a beacon of modern education.

- *AI in language learning and content creation*: AI's role in language learning, as seen in platforms like Duolingo, demonstrates its potential to make education both effective and enjoyable. AI-powered chatbots offer interactive language learning, making foreign languages more accessible. In content creation, AI tools like Brainly act as homework helpers, adding an element of fun to the learning process.
- *Visualization and interactive learning*: AI tools and VR and AR technologies redefine how complex concepts are understood and taught. Tools like Genius 3D Learning and Wolfram Alpha enable students to visualize mathematical concepts and complex data in interactive ways, enhancing comprehension and engagement.
- *Predictive analytics*: AI will analyze student data (within privacy guidelines) to predict future performance and identify students at risk of failing. This allows for early intervention and personalized support, improving overall academic outcomes.
- *AI-powered content creation*: AI can generate personalized learning materials, such as customized practice problems, interactive simulations, and even personalized textbooks. This can free up educators' time and ensure students have access to relevant and engaging resources.
- *24/7 intelligent support*: AI-powered chatbots and virtual assistants can answer student questions instantly and offer academic support whenever it is needed. This can be especially helpful for students needing additional clarification or studying outside of regular school hours.
- *Focus on emotional well-being*: AI might integrate emotion recognition technology to understand student moods and adjust teaching approaches accordingly. This can create a more supportive learning environment and address potential mental health concerns.
- *Beyond the classroom*: AI can transform professional development for educators, personalize career guidance for students, and even create AI-powered learning companions that motivate and support students throughout their learning journey.

The potential of AI is exciting to consider, but its success hinges on a partnership between technology and humanity. Successful implementation of AI in education includes addressing biases, embedding ethics into algorithms, and nurturing creativity and social-emotional skills. True innovation requires collaboration: educators help the AI with their understanding of the human learner, and the AI performs the mundane classroom tasks so teachers can interact more with the students. AI in education offers a chance to not only rethink how we teach but to fundamentally enhance the joy of lifelong learning. AI is a tool that can be shaped by educators to provide inclusive and ethical lessons that promote a love of learning and students' curiosity.

AI IN FINANCE

Finance relies on information and can benefit from technological innovation. Technology has fundamentally reshaped how we understand, trade, and manage money, from the first ticker tape machines to computerized stock exchanges. AI can revolutionize how financial institutions operate and individuals interact with their finances.

AI's ability to process vast amounts of data, identify patterns, and make autonomous decisions is promising for the financial sector. AI may be able to make sense of market signals, news feeds, customer information, and regulatory updates. It can make investments more accessible, risk management more effective, and financial services more inclusive and responsive to consumer needs.

There is a need for careful, intentional adoption. Using AI in finance must incorporate the theme of "shaping a better future through responsible innovation and human collaboration." Our aim is to show how AI can be of benefit in finance while emphasizing the ethical considerations, the necessity of human oversight, and the collaborative potential. With strategic integration, careful development, and the use of AI, progress can be made in the area of finance.

AI's Influence on Finance

Finance builds economies, empowers individuals, and can (when responsibly deployed) motivate positive change. AI has the potential to reshape financial access, expanding opportunities for historically underserved communities. This powerful tool demands thoughtful deployment, though. We must address its limitations, potential biases, and the necessity of a human-AI partnership to unlock its positive potential within an ethical, transparent framework.

Robo-Advisers and the Democratization of Wealth Management

AI-powered robo-advice platforms are changing how people get advice in the wealth management industry. Using sophisticated algorithms and machine learning capabilities, these platforms automate core aspects of financial planning, including portfolio creation, rebalancing, and tax optimization. By automating traditionally labor-intensive processes, robo-advisers deliver tailored financial advice, once reserved for the wealthy, through cost-effective and scalable service models.

Robo-advisers are accessible and convenient. Individuals who could not previously afford the fees of human advisers can now receive personalized investment guidance and automated portfolio management at a low cost. The user-friendly interfaces and simplified approach attract a broad consumer base, especially younger investors and those comfortable with technology-driven solutions.

We should emphasize the responsible implementation of such types of financial AI to ensure they do not cause problems for its users. Robo-advisers use data to optimize investment decisions based on a user's financial goals and risk profile. There are limitations to providing advice during emotionally charged scenarios, like economic recessions, sudden life changes, or unexpected market volatility. These events have profound real-world consequences on financial plans. The future of wealth management likely lies in blending human-AI collaboration: combining the strengths of financial data analysis with a human understanding of sound financial decision-making.

Algorithmic Trading and New Market Dynamics

The high-stakes, fast-paced world of trading is an arena where AI has the potential to be profoundly transformative. Hedge funds and trading firms increasingly rely on machine learning algorithms to analyze market data, identify subtle signals and patterns, and predict trends. These AI-powered systems can execute trades at speeds and volumes inconceivable to human traders, potentially offering a significant edge in markets where milliseconds matter.

The integration of algorithmic trading introduces some important considerations for ensuring the stability of financial markets. While offering new profit opportunities, AI-driven trading can also amplify the potential for market disruptions and instability. Regulatory oversight and safeguards are needed to prevent events like *flash crashes*, where algorithms can unintentionally cascade, causing problems within markets. Transparency and human overseers are essential as companies increasingly rely on complex AI models for financial purposes.

Risk Management in the Age of Big Data

Businesses must effectively predict and mitigate risk within the financial sector. AI can assist with these activities because of its ability to manage large amounts of structured and unstructured data. Machine learning models are used by banks and financial institutions to manage data, uncover insights, and build sophisticated risk assessment profiles. AI assistants can create more reliable early warning systems, identifying potential market instabilities or credit risks and offering invaluable foresight for preventive actions.

AI in risk management is about more than just data analysis; it is about making proactive, data-informed decisions. AI systems excel at recognizing patterns that might remain unseen by human analysts. AI enables institutions to respond to emerging trends and implement mitigating strategies by detecting subtle risk factors in loan portfolios, anticipating fluctuations in asset values, or identifying problems within complex financial transactions. These capabilities help businesses make informed lending decisions, optimize their portfolio allocations, and create more resilient business models.

Identifying Fraud and Bolstering Cybersecurity

Due to financial cybercrime, companies need adaptive, intelligent security solutions. AI can be used as a kind of "digital guardian," analyzing enormous volumes of financial transactions to detect inconsistencies that could indicate fraudulent activities. Anomaly detection, fueled by advanced machine learning techniques, provides a first line of defense. Real-time monitoring powered by predictive analytics proactively identifies vulnerable endpoints and patterns indicative of fraudulent behavior.

While protecting existing systems, AI also bolsters cybersecurity through predictive threat modeling and simulation. These AI-trained models replicate many real-world attack scenarios on critical financial infrastructures, identifying potential weaknesses and proactively developing preventive solutions. AI-based defense mechanisms have the potential to outsmart adversaries and build adaptive systems that learn from previous attack models.

Redefining Creditworthiness and Financial Inclusion

Traditional credit scoring systems, which rely on credit histories, can fail to successfully identify an individual's creditworthiness. This can result in an entire economic class being denied loans and financial services based on limited information, further contributing to a cycle of financial exclusion. AI can overcome this challenge by analyzing alternative data sources that more accurately reflect financial stability and responsible behaviors.

Integrating unconventional data points for credit evaluations can make the process of identifying credit worthiness more inclusive. Consider utility payment histories, rental payment records, and even responsible social media usage as contributing to more dynamic, nuanced creditworthiness profiles. A wider variety of individuals can be deemed creditworthy when using machine learning models. These models excel at evaluating vast datasets and increasing access to financial products like loans, mortgages, and financial tools that promote economic growth and financial stability for historically underserved communities.

AI and Personalization

AI in finance is not simply a new form of automation; it is a significant change in how we interact with our financial lives. The future of AI includes far more than robo-advisers or more inclusive credit assessments. It offers the potential of *hyper-personalized* (or uniquely personalized) finance and a new role for human expertise.

AI as Financial Companion

Let's consider another future scenario: Imagine that an AI knows your bank balance, aspirations, and anxieties. This new kind of AI financial tool does not just process transactions. It also analyzes your spending habits, identifies potential pitfalls, and suggests specific financial products that address your unique needs. Insurance plans created by this AI adapt to your lifestyle in real time. In our future scenario, AI chatbots can help with your account balance inquiries, and also offer financial guidance that helps you meet your financial goals. This hyper-personalization helps individuals better understand and proactively manage their financial well-being.

While the AI financial assistant in our future scenario has incredible functionality, the financial relationship is largely about assisting the person. True innovation rests in empowering AI-driven solutions alongside reimagined human interaction. For wealth managers and financial advisers, AI can become a potent tool to help them perform their work. Data-driven insights inform recommendations, routine tasks are automated, and financial managers obtain valuable time to guide clients through the emotional complexities that go hand-in-hand with major financial decisions. Advisers manage portfolios, but they also address the emotions that go along with legacy planning. They can help their clients better manage their emotional reactions to economic downswings or guide clients through major life choices like homeownership.

Workforce Evolution: Reskilling Because of AI

Implementing AI solutions in many industries creates new efficiencies and personalized services, and so managing its impact on the workforce is vital. AI streamlines repetitive tasks, potentially automating positions involving routine data analysis or data entry. People must proactively adapt to these changes. *Reskilling*, where already skilled employees are trained in different types of work, can prepare individuals for new roles caused by the use of AI. Some of these skills will include those needed to work on AI. For example, more financial data scientists will be needed to build the systems and analyze and interpret their output. AI ethicists will be required to ensure algorithmic fairness, while strategists specializing in human-AI collaboration will optimize how AI is deployed and managed.

Imperatives for Ethical and Sustainable AI in Finance

AI's true transformative potential remains inseparable from addressing ethical challenges and the critical importance of human oversight. Algorithmic bias must be addressed early in the AI system development. Data auditing must be performed to eliminate unfairness in training sets. Diverse development teams and strategies can prevent *model bias*, which refers to unintentionally encoding harmful prejudices into the systems meant to provide neutral assessments. Explainability frameworks like XAI are important in creating trust in the AI systems, and they safeguard both the business interests and consumers' interests.

Regulation should be seen as a "roadmap" to promote algorithmic transparency and ensure explainability. This means that businesses will have to manage their need to protect trade secrets while maintaining the trust of their customers. AI models should be open to human review, and businesses should provide a clear explanation of how and why financial recommendations are generated. Responsible companies have a vested interest in building trust with their consumers.

Within financial institutions, governance of AI-based systems needs clear accountability structures and a sense of collective responsibility. The future of AI in finance requires a united approach of multiple stakeholders, including those on the board of the company and the technology department. Collaborations among tech developers, banks, policymakers, and consumer advocacy groups will help AI's ethical development and promote a broader understanding of its transformative impact.

A Better Future with AI Collaboration

AI has the potential to "democratize" financial services, optimize investment strategies, and combat the threats of fraud and instability. However, it is a tool that gains its real power only when we use it with intent, foresight, and an unyielding commitment to ethics. AI will continue to redefine the financial services industry, but people need to consider the ethical implications of its use. Only through intentional collaboration, a focus on inclusion, and the recognition that truly sustainable progress rests on an alliance between algorithm and empathy can we build a future of finance that reflects the best of human and machine innovation.

AI in Finance Case Studies

Case Study 1: An innovative fintech lending platform

The Problem: Traditional credit score systems unintentionally keep certain members of society out of the formal financial system. Individuals like recent graduates, immigrants, or those with non-traditional careers may be financially responsible but can be denied loans due to the lack of a credit history.

AI Solution: The platform deploys machine learning algorithms to search alternative data sources. They might analyze rental payment histories, utility bill consistency, educational attainment, and even social media behaviors to establish a holistic picture of the individual's creditworthiness.

Impact: Using AI to identify positive patterns outside of regular credit reporting, the platform extends loans to people excluded by the conventional system. This gives capital to small businesses who need it to expand. It can also further people's education, help them responsibly manage debt, and increase financial security.

Case Study 2: A leading hedge fund specializing in high-frequency trading

The Problem: Human traders cannot analyze enormous datasets or react quickly enough to capitalize on many market opportunities consistently. Profitability increasingly depends on technological supremacy over competitors.

AI Solution: The hedge fund uses sophisticated algorithms for high-frequency trading, seeking patterns in everything from financial news sentiment to order book activity and global exchange data. Machine learning models trained on historical market data identify subtle signals often missed by humans, predicting price movements for specific stocks, commodities, or currencies. The models trigger and execute trades autonomously within milliseconds.

Impact: The use of AI profoundly alters market dynamics. Successful high-frequency trading can yield significant profits for the hedge fund but also requires stringent regulation to prevent systemic risk and safeguard fair markets. The importance of algorithm transparency and human oversight intensifies within this volatile space.

Case Study 3: A major international bank

The Problem: As the financial sector implements more digital solutions, it must manage increasing cyberattacks targeting internal systems and consumer data. Security breaches are expensive and severely damage consumer trust.

AI Solution: The bank implements AI-powered cybersecurity systems. Behavioral analytics identifies anomalies in transaction logs, such as suspicious transfers and unusual geographic locations. Predictive modeling simulates potential attack scenarios, proactively uncovering vulnerabilities before exploiting them. Through machine learning, AI defenses continuously evolve to match the ingenuity of cybercriminals.

Impact: With AI augmenting traditional cybersecurity mechanisms, the bank reduces data breaches and fraudulent activity, safeguarding billions of dollars in client assets. This success enhances financial security for consumers and promotes confidence in digital banking operations.

Case Study 4: An insurance start-up using IoT devices and machine learning

The Problem: The insurance industry remains risk-averse and slow to change. Generalized parameters determine premiums and claims adjusters primarily deal with incidents.

AI Solution: The start-up introduces Internet of Things (IoT) sensors (such as fitness trackers and smart home devices) as part of its services. Individuals can choose to use the devices, understanding that AI analysis could lead to adjusted premiums or personalized coverage tailored to their lifestyle. Insurers might reward customers with demonstrated healthy behaviors based on verified data or suggest preventive measures.

Impact: More equitable, personalized insurance models emerge. The potential exists for insurers to incentivize safer driving, promote fire prevention with discounts, or create a system where low-risk clients truly feel rewarded. Data privacy becomes a significant issue for both consumers and industry.

AI IN MANUFACTURING

Manufacturing, the act of creation on an industrial scale, has defined eras. From the steam engine to the assembly line, ingenuity has helped people craft and produce the goods shaping our world. Today, AI is influencing manufacturing in significant ways that require careful consideration and an understanding of ethics.

In the future, machines do not merely follow instructions; they learn from large amounts of data, adapting to evolving realities on the factory floor. Linear schedules and rigid processes will be replaced with flexible, AI-developed production lines that meet real-time demand. Algorithms analyze sensor feedback to predict breakdowns before they occur, safeguarding productivity and optimizing maintenance. Computer vision systems with superhuman precision ensure zero-defect parts and drive continuous quality improvement. AI in manufacturing can optimize operations and change how problems are identified and solutions are conceived.

There is a complex interplay of human and machine that occurs in manufacturing. As production lines become autonomous, new challenges emerge. How do we harness AI's relentless efficiency without sacrificing the human judgment developed over years of on-the-floor experience? Can AI models trained on large amounts of data genuinely apply the "gut instincts" of an experienced engineer? How do we train AI-savvy manufacturing professionals ready to collaborate with sophisticated algorithms to make better decisions?

This section examines the influence of artificial intelligence in manufacturing. We consider AI-powered transformations reshaping production from shop floor to supply chain, but the focus extends beyond technology alone. We aim to examine ethical considerations of AI, workforce evolution, and the pursuit of sustainable, adaptive manufacturing systems. We will learn about more than "smart factories." The future of manufacturing is intelligent, interconnected, and constant learning. Artificial intelligence will change how things are made.

Mass Customization and the Consumer Connection

The age of generic, "one-size-fits-all" mass production is waning. The future consumer will want to be involved in the production process. AI will assist this process by analyzing large amounts of data about consumer preferences, lifestyles, and biometric data. The use of AI means that consumers will be able to express themselves through manufactured goods.

AI does not only tailor sizes and colors of products. It can also predict unconscious desires. Generative AI models, trained on enormous datasets of aesthetics and functional properties, can propose novel product variations in response to consumer data. It could generate the "perfect" running shoe, which has the ideal cushioning for your stride, the right style for your wardrobe, and perhaps even materials sourced ethically in line with your values.

This AI-driven approach is more like a collaboration between the consumer and algorithm than that of traditional shopping.

The result of implementing AI to produce goods for specific consumers may result in significant changes for manufacturers. Small manufacturers would have an advantage over their larger rivals. Their ability to change quickly becomes an asset. AI-optimized small-batch production lets them quickly respond to emerging trends and collaborate with consumer groups to generate niche products on demand. This approach could result in custom-designed sporting equipment built for an athlete's unique physique, medical devices calibrated to patient anatomy, or fashion that is customized for the individuality of its wearer.

When Algorithms Create and Manage Systems

The "factory of tomorrow" will be more akin to a living ecosystem than an assembly line. Let us imagine how AI can change the future of factories: Imagine interconnected AI systems making materials, generating energy, and managing information flows to ensure that the entire process is optimized. This change in manufacturing will alter every part of the industry.

- *There is a new relationship between the physical and digital parts of the process.* The digital twin becomes commonplace: AI maintains a detailed, real-time model of the physical production facility. This allows companies to run numerous simulations, identifying operational bottlenecks before they appear on the factory floor, testing production changes within the digital environment, and helping with planning and development.
- *A circular and sustainable way of manufacturing is created.* The future of AI in manufacturing will involve conscious production. AI will manage resources, analyzing real-time energy use, materials wastage, and even the potential downstream environmental impact of production choices. True sustainability does not simply reduce harm; it is also about designing goods for easy reuse, recycling, or upcycling. AI helps achieve this throughout a product's entire life cycle.
- *Anyone can become a designer.* Generative AI can enhance 3D printing and additive manufacturing. AI-generated designs that mimic nature's efficiency and optimize for minimal material use can create processes and products better than those we have today. We might see aircraft components resembling birds' bones – strong yet impossibly lightweight. AI can help people with new ways of thinking about form and function.

From Factory Floor to Data Center

Future factory workers may not wear overalls: they may use tablets and data-sets. Dismissing the implementation of AI as pure automation is shortsighted. AI augments human skills. There is power in humans using AI to engage in superior decision-making.

◾ *"Data fluency" becomes a universal skill.* Manufacturing will rely on blueprints and specifications, as well as code and statistical models. Basic programming concepts and AI literacy will become as basic a skill as using a wrench on an assembly line. Workers who can interact with machine learning systems, adjust outputs, and help refine models become critical to maximizing the potential of AI-driven factories.

◾ *The new role of manufacturing data artist will be created.* AI produces data, but understanding the nuances of those insights requires unique abilities. Skilled workers help to visualize complex datasets, identifying hidden correlations and anomalies. They use their knowledge of how things should work combined with AI-highlighted patterns leading to swift diagnosis and solution optimization.

◾ *AI trainers and ethical "guardians" will be needed.* While some jobs become obsolete, new roles with far-reaching consequences emerge. Workers will teach AI about the realities of production, such as what a perfect weld looks like and how vibrations differ when there is wear on a machine. They will train the AI in their experience and knowledge. Additionally, monitoring for unintended model bias will become an essential workforce component, preventing discriminatory outcomes from being hidden in complex formulas.

AI in Manufacturing Case Studies

Case Study 1: An innovative start-up specializing in 3D-printed custom parts

The Company: ProShape Ltd. is a small start-up specializing in on-demand, 3D-printed prosthetic limbs, and medical devices.

The Problem: Traditional prototyping and fabrication of prosthetics is labor-intensive, costly, and offers limited customization. This creates access barriers, especially for children who quickly outgrow prosthetics, creating financial and psychological hardship.

AI Solution: ProShape employs several aspects of AI to create the next generation of custom-fit medical devices:

◾ *Biometric scan analysis:* Patients are scanned by advanced imaging systems. AI transforms these into highly accurate 3D models for tailored design.

◾ *Generative AI design:* Algorithms analyze the user's anatomy, mobility needs, and even aesthetic preferences to propose multiple prosthetic designs optimized for performance and comfort.

◾ *Automated 3D printing workflows:* Machine learning selects ideal materials for printer settings and even optimizes the printing process itself for maximum strength and flexibility without bulky constructions.

Impact: ProShape improves patient care. Affordability, rapid adaptation, and bespoke personalization transform lives and improve access to prosthetics. Additionally, ProShape leverages this platform to provide prosthetic devices to NGOs working in resource-limited regions, demonstrating AI's ability to scale manufacturing for social good.

Case Study 2: A large automotive company using AI-powered predictive maintenance for its manufacturing facilities

The Company: A major global automaker with several vast, highly automated assembly plants worldwide.

The Problem: Unplanned downtime can cripple assembly lines, causing delays cascading through the production schedule and costing millions per hour.

AI Solution: An extensive network of sensors collects data on every stage of the assembly process: vibrations, temperature changes, motor power consumption, and even audio emissions from manufacturing equipment. AI models analyze this data in real time to establish machine health baselines and detect the subtlest deviations that signal a potential upcoming failure. This enables

- *Predictive scheduling:* Maintenance is scheduled only when needed, maximizing machine uptime and eliminating wasteful replacement of still-functioning parts.
- *Preemptive parts ordering:* The system can even predict which specific components will likely fail soon, streamlining their supply to maximize productivity.

Impact: Production downtime declines dramatically, ensuring delivery schedules are met and maximizing production capacity. The automaker realizes significant savings and gains confidence in consistently meeting delivery commitments to clients.

Case Study 3: A multinational electronics manufacturer deploying AI in global supply chain risk management

The Company: A multinational giant producing smartphones, components, and a range of consumer electronics.

The Problem: Global supply chains are complex, with disruptions having long-reaching effects. Natural disasters, trade wars, and geopolitical instability threaten materials sourcing and timely delivery.

AI Solution: The AI system ingests immense amounts of data, including:

- real-time shipping tracker data, port congestion info, and global transportation updates
- news feeds worldwide in multiple languages, focusing on geopolitical risks and emerging conflicts
- social media sentiment analysis to identify potential unrest or labor issues
- raw materials prices, commodity trends, and real-time exchange rates

Impact: AI becomes a predictive risk analysis tool. Its models continuously scan for potential vulnerabilities in the supply chain. It triggers alerts, recommends hedging strategies, identifies alternative procurement routes, and suggests temporary inventory build-ups in certain zones to maintain production output.

Case Study 4: A furniture manufacturer implementing AI-driven demand forecasting and dynamic production planning models

The Company: A mid-sized, bespoke furniture manufacturer specializing in crafted wooden pieces for homes and luxury hotels.

The Problem: Fluctuating demands on production, lack of trend predictions, raw materials procurement inconsistencies, and lead times made the existing systems unreliable.

AI Solution: The company deploys AI at multiple stages:

- *design trend AI:* Algorithms analyze social media trends, social media posts, influencer blogs, and architectural design forums to anticipate consumer desire for colors, styles, and sustainable materials.
- *predictive demand forecasting:* Machine learning analyzes historical sales data alongside socioeconomic indicators, seasonal trends, and regional weather patterns to anticipate surges and potential slowdowns.

Impact: The company implements "just-in-time" manufacturing principles for maximum efficiency. Raw materials procurement aligns with expected demand, minimizing wastage and warehouse carrying costs. The craftspeople focus on their strengths, while AI optimizes the logistics.

AI IN ENTERTAINMENT

Art and technology have always been a part of the entertainment industry. People used to gather around campfires for enthralling tales, and now we listen to music on CDs and appreciate movies on theater screens. With each technological advancement, our entertainment changes, becoming more engaging, interactive, and less limited. AI will allow us to become participants within entertainment.

This section examines how AI affects entertainment, from scripting and creation to the personalized recommendations served on your streaming service. Technology alone cannot dictate the future of entertainment. Just as we did with other technological advancements, we must carefully consider AI's influence on creativity, the potential for new biases, and how human artists and technologists use it in their work.

AI and Immersive Storytelling

Who are your favorite fictional characters? What makes them compelling? Complex choices affect how storylines are developed. AI is a powerful tool for storytellers: it helps them to manage this complexity, generating richer narratives and branching possibilities. AI will also redefine the roles of scriptwriters and game designers.

Character Creation

AI models trained on extensive databases of literature and human behavioral data can develop complex characters. These AI-generated personas will have flaws, quirks, and their own motivations that influence the narrative arc. AI could become a screenwriter's writing partner, proposing plot twists and unexpected character development based on logic.

Adaptive Narratives

From children's picture books to fantasy video games, interactive storytelling will increase in sophistication. AI can develop dynamic stories, responding to a player's choices, mood, and physiological signals (captured with consent from wearable sensors). Because the reader or listener can participate in the story, AI can foster a sense of being a part of the story.

Worldbuilding with AI

Crafting compelling environments, whether fantasy kingdoms or historical dramas, requires the utilization of meticulous detail. AI can easily generate these types of details to create novel experiences or analyze past eras to ensure scripts are free from inaccuracies. This allows writers to focus on the characters' emotional journeys, knowing the backdrop aligns with the plot of the story.

AI-Generated Music

Algorithms can learn musical genres, predict melodies that evoke specific emotions, and generate entire soundtracks personalized to our preferences. This challenges the notion of music being a "pure expression" of a lone artist. Can we still experience emotion from a symphony written collaboratively by human and machine?

Deepfakes and Synthetic Voices

The ability to fabricate realistic videos and mimic human voices makes AI a unique tool to generate entertainment. This ability, however, raises some ethical concerns about its use. An AI copying the voice or image of deceased actors may delight some people but may be seen as exploitation by others. Using AI voice "clones" of actual human actors could threaten the role of professional voice actors.

The "Democratization" of Entertainment

AI can make both creation and consumption more inclusive.

Bespoke Digital Experiences

AI-powered game mechanics can adapt to individual players' abilities and generate unique VR environments that address the user's physical limitations. AI can create an inclusive entertainment experience that can accommodate disabilities and make "fun" universally accessible.

Translation

AI can provide real-time dubbing and subtitling for a variety of languages. This can allow stories from around the globe to reach larger audiences.

The On-Demand Audience Analyst

Entertainment studios rely on feedback. AI can offer sentiment analysis capabilities for a range of feedback services, from preview viewers to live Twitch reaction data. As a "data analyst," AI will influence decisions, potentially homogenizing content for mainstream appeal. Will creative risk-taking survive an algorithmic focus group?

AI's Broader Impact

With any new technology, thoughtful implementation outweighs blind enthusiasm. Consider the influence of AI in entertainment in these ways, as well:

- *Identifying algorithmic bias in creation.* AI models trained on existing works inevitably risk replicating the biases present within them. Could an AI "scriptwriter" trained primarily on male-authored dramas subtly favor male protagonists? Ethical oversight and bias audits of datasets are crucial to prevent the perpetuation of these inequalities in AI-generated content.
- *"Democratization" vs. "dilution" of creative work.* Could AI tools make anyone a filmmaker, storyteller, or composer? It raises questions about access to art but concerns over-saturation and dilutes the value of genuine creative skill honed over a lifetime.
- *Redefining fandom and the creator-audience relationship.* As AI-created storylines become interactive, will new online communities form around shared narrative experiences? Could we enter an era of personalized endings or "fansourced" content direction influencing major film releases? This could cause significant intellectual property challenges for a changing industry.

The Future of AI-Augmented Entertainment

There will be few many changes once AI is widely implemented in entertainment. AI-driven entertainment's most compelling experiences will appeal to all of our senses and encourage us to participate, collaborate, and discover new things.

The End of Passive Viewership

The experience of sitting on a sofa to watch shows may become obsolete. AI's influence will change the audience from spectators into active co-creators. Consider the following ideas:

- *"Holodecks" on demand.* We will no longer sit and passively view shows. Haptic suits, spatial tracking, and AI-generated worlds will be used during the program, and our own movements and preferences will alter the experience of the show. For example, if you are feeling stressed, the AI can create a martial art quest you can complete inside your living room. If you feel like you would like to explore, the AI can change your apartment into an interactive Martian outpost, teeming with computer-generated flora and fauna to investigate.

- *"Live scripted" events.* AI will change the way reality TV operates. There will be no more staged scenarios. These will be replaced by immersive events driven by AI. The participants' emotional responses, choices, and interactions will shape reality drama with surprising twists engineered by the AI system that analyzes individual motivations and group dynamics in real time. This type of entertainment, however, does show how ethical problems can arise in entertainment. It also can potentially reveal aspects of human behavior under the extraordinary circumstances AI subtly shapes.

AI-Powered Personalization

Algorithmic recommendations are aimed at keeping people interested in watching shows. In the future, AI will make entertainment into an art form that responds to our feelings.

- *Music that heals.* AI, equipped with a considerable information about music theory and its effect on human physiology, will craft songs that adapt to a person's feelings in real time. It could provide a musical selection to soothe a person's feelings after a hard day at work, use specific harmonizing minor chords to help a person safely process grief, or generate an energizing song to enhance a workout. The AI assistant can provide custom sounds that improve our personal lives.

- *Dreams as movies.* With sleep analysis tools, AI will learn to analyze our subconscious visual language. Upon waking, dream snippets can be then transformed into abstract animations or surreal storylines. Instead of journaling in the morning, we can watch videos of our dreams curated by AI, perhaps finding profound meaning in them through AI's artistic reinterpretations of our own dreams.

The Problem of Hyper-Personalization

The idea of hyper-personalization using AI raises valid ethical concerns. Does hyper-personalization result in people becoming too self-centered? Will AI-created entertainment prevent people from going outside for a walk? These are critical conversations we must have as a society. AI in entertainment has the potential to go far beyond merely replacing artists. Here are some ways that AI may improve our entertainment:

- *Unleashing latent creativity:* AI tools can enable anyone to visualize ideas, iterate rapidly on art, or orchestrate their own interactive stories will not eliminate established talents but provide pathways for previously unknown artists. AI could assist authors with dyslexia so they can share their stories or an elderly veteran can use generative AI to turn war memories into paintings.
- *Preservation and reimagination:* Future artists will not erase the artistic achievements of those from the past. AI will become a steward, restoring decaying films, revising beloved games for new generations, and helping linguists decipher ancient scripts to reveal performance works.

The future of AI entertainment contains unprecedented levels of immersion, personalization, and the opportunity to unlock our shared and individual creative potential. Our task now is not merely to anticipate these transformations; it is to participate in shaping the ethical, inclusive, and truly awe-inspiring experiences AI helps create. After all, the best stories are written together.

AI in Entertainment Case Studies

Case Study 1: A pioneering immersive entertainment start-up

Problem: While VR and AR headsets exist, experiences often feel disjointed or uncomfortable, failing to deliver on the promise of true immersion.

AI Solution: The start-up has a multifaceted approach:

- *Neuro-adaptive storytelling:* Wearable sensors monitor the player's physiological responses (such as skin conductivity for excitement and gaze tracking for focus) to shape the experience in real time. A horror game becomes more frightening if you are genuinely terrified or less so if you are overwhelmed.
- *AI-generated backdrops:* Detailed worlds can be created around a player, which can react to their movement and choices. The game environment can transform as the player plays the game, enhancing their sense of presence within the virtual world.

Impact: Entertainment experiences become adaptive and create a stronger connection between the player and the virtual world.

Case Study 2: A major music streaming platform

Problem: Recommendations have become predictable. Users desire more than just the same genres or artists.

AI Solution: The company has invested in advanced affective AI:

- *Mood sensing:* AI analyzes what users play, as well as when and under what circumstances. Are they compiling an eclectic party playlist or melancholic late-night songs? The choices inform its suggestions.
- *Other inputs:* Connected devices (with user permission) could give the AI additional insight. If the user is reading exciting material, the AI then may suggest energetic music. The user's slow, meditative breathing patterns signal a need for ambient, calming sounds.

Impact: The AI platform not only can choose more appropriate music, but it can affect the user's moods. Users develop a stronger connection with the chosen music as a reflection of their personal lives, which may help them better understand themselves.

Case Study 3: An independent film studio specializing in interactive narratives

Problem: Interactive, branching-narrative films are often poorly developed, with decision points breaking the feeling of immersion in the story.

AI Solution: The studio deploys sophisticated natural language processing (NLP) tools:

- *AI character personas:* Instead of only following certain decision trees, characters can have their own AI engine. Responses change according to the player's tone, language choices, and past interactions with them.
- *Adaptive scoring:* Music is procedural, shifting in tension and style to heighten moments of connection or build suspense based on the narrative's path.

Impact: Players enter a truly "conversational" relationship with the narrative. Instead of making overt game-like choices, characters react believably to their input. Emotional impact and the ability to replay a game are altered.

Case Study 4: A global museum consortium

Problem: Preserving endangered cultural heritage (music, dance, and performance art) is usually done in traditional display cases. Visitors can get an idea about the work, but not a full understanding of it.

AI Solution: The consortium embarks on an ambitious, collaborative project:

- *Generative dance reconstruction:* AI models trained on archived text descriptions, partial visual records, and biomechanics are used to recreate lost dance styles. Visitors can then learn through imitation, with their motions being compared by AI to provide virtual lessons.
- *Lost soundscapes:* AI creates sound recordings to complete fragmented musical scores, reconstruct ancient instruments' sound profiles, and generate the ambiance of lost performance spaces based on their ruins.

Impact: Museum exhibits become more interactive. History can become more realistic to the museum visitors who can see and engage with the interactive exhibits. AI assists in bridging the differences between cultures so that people from different places can better understand art forms at risk of disappearing from memory.

AI IN THE ENVIRONMENT

Climate change is a fundamental restructuring of our planet's natural rhythms. Traditional human analysis tools cannot fully measure natural systems' complexity. AI can help us understand the effects of climate change. It can easily manage large datasets of climate information, uncovering previously unforeseen connections and allowing scientists to make better predictions. These models do not simply react to climate trends; they help us anticipate future climate issues.

AI-driven prediction models allow us to better understand the effects of climate change on sea levels and could help us identify which cities are most vulnerable, pinpoint coastal regions where agricultural collapse is likely, and provide governments with the insights needed to enact targeted, anticipatory policies well before disaster strikes. AI can help us not only track changes, but also plan for them. Additionally, this approach unlocks the potential for better local climate forecasts. For example, AI models specifically tailored to a major agricultural region can alert farmers to subtle shifts in precipitation or optimal crop selection within long-term trends. These fine-grained insights allow people to better adapt to the changing climate, not just on a global scale, but also for those individuals most at risk from its cascading consequences.

AI can change our perception of natural cycles. Let us consider an AI model trained on the last century of climate data, as well as on paleoclimate data extracted from ice cores and geological records going back millennia. This data may reveal long-term oscillations in the weather that could significantly enhance our understanding of current warming trends. We could use these insights to answer whether current conditions are unprecedented or part of a longer cycle, which could assist research and the formulation of public policy. Moreover, it raises fascinating philosophical questions. Is the "ideal" state for the Earth's climate the world experienced by pre-industrial ancestors?

Naturally, it is essential to be discerning, as any predictive model relies on the data it is fed. Here, human-AI collaboration remains vital. Environmental scientists bring domain-specific knowledge to ensure the AI models are not making flawed assumptions. They understand the physics and chemistry governing these systems beyond what an algorithm can infer purely from statistics. It is about understanding that AI models are powerful but fallible representations of complex reality. The true potential of AI lies in these systems informing conversations about mitigation strategies, adaptation planning, and fostering deeper awareness of the consequences of our ecological choices.

AI as the "Steward" of Natural Resources

Utilizing natural materials and energy and increasing production has allowed human progress, but there has been an ecological cost. AI can help us intelligently optimize the use of our natural resources. It can assist us in decoupling economic well-being from unchecked resource depletion and environmental destruction.

Consider a future where "smart grids" powered by AI become common. Smart grids balance fluctuating energy sources and intelligently adapt to individual consumer profiles in real time. AI could incentivize less critical usage during peak loads by offering temporary discounts on utility bills while ensuring that hospitals and critical infrastructure are prioritized. As a result, consumers have a more stable energy grid and energy becomes a commodity.

Sustainability is an important issue that can be addressed by AI. AI has the potential to address pollution reduction through innovative closed-loop production cycles. Advanced recycling facilities could use AI-powered image recognition algorithms to sort refuse far more efficiently and precisely than ever before. Instead of simply burning or burying garbage, certain materials could be diverted to manufacturers for reuse. AI will not instantly create a zero-waste society, but it could dramatically boost material recovery rates and create a situation where materials retain their intrinsic value for far longer than they do now.

Of course, AI's role must expand beyond managing what we create to informing how we develop it in the first instance. Generative AI models can be utilized to enhance responsible use. In collaboration with material scientists, such systems could explore infinite potential material combinations for more efficient batteries, biodegradable plastics, or solar cells with improved properties. AI would avoid the need for resources for wasteful trial-and-error prototyping.

AI Can Manage the Biosphere

Biodiversity is important to planetary resilience. Our forests, oceans, and all their non-human inhabitants are part of a natural balance and act as a buffer against catastrophic ecological breakdowns. Yet biodiversity loss's scale, speed, and remoteness hinder our ability to intervene effectively. AI can serve to extend our ability to measure and analyze.

The oceans are also vital ecosystems under threat. AI-powered acoustic tracking systems can monitor unique whale songs, tracing their populations with greater detail than ever before. These patterns can become indicators of habitat health, shipping lane threats, and the impact of industrial noise pollution on the communication methods between species. Additionally, underwater drones fitted with AI-trained imaging systems can classify coral reef biodiversity, identifying early signs of bleaching. AI-powered systems can establish baseline health and potential ecological threat warnings on a planetary scale.

AI can effectively utilize satellite images. These types of systems can identify early markers of illegal deforestation in dense rainforests. These systems may also alert park rangers or concerned NGOs, providing timely, actionable data instead of a hindsight view when the rainforest destruction is complete. Similarly, AI-driven drones can be used to analyze wildfire risks, model fuel loads in dry forests, monitor human encroachment patterns, and help agencies proactively manage them through pre-emptive controlled burns and strategic infrastructure.

With this large amount of data and analytical power comes responsibility. Conservation is not just about preserving individual species. AI-powered modeling reveals complex dependencies within habitats, showing that saving a single animal may be pointless if the local flora or insect populations it relies upon are already going extinct. These insights will likely challenge the way we currently care for the planet and create complex moral dilemmas. Which ecosystem does an AI system prioritize when there are not enough resources to safeguard all? It is here that ethics and transparency around algorithm development become paramount.

Prioritizing Action with Conflicting AI Predictions

The scale and interconnected nature of Earth's climate systems create inherent uncertainty, even within scientific modeling. AI offers unprecedented analytical power to process climate data but cannot entirely remove uncertainty. Different models built on slightly varying data, even those employing cutting-edge techniques, will occasionally present conflicting scenarios about the severity and time scale of projected impacts. This leaves policymakers in a tremendously difficult position. Acting early based on the direst scenario might avert catastrophe but could also prove economically disruptive if those outcomes are later shown to be overestimated. Cautious optimism and waiting for greater consensus can become dangerous complacency if AI warnings about potential "tipping points" prove accurate.

It is important to understand how AI differs from previous climate modeling approaches. Machine learning systems might uncover overlooked correlations, such as how subtle shifts in a certain deep-sea current influence rainfall patterns thousands of miles away, an outcome conventional models have not even been programmed to look for. However, does this mean unquestioningly trusting a data-driven relationship between scientists with expertise in the underlying physics driving these climatic events? AI is not an omniscient entity but a powerful pattern detector. Striking a balance between its insights and established domain-specific knowledge is crucial to avoid catastrophic decision-making failures.

The problem deepens when this predictive ability is affected by political agendas and the influence of lobby groups. AI can then become a "weapon" in the debate. Factions seeking inaction can highlight any divergence of opinion among climate models as proof that the "science is uncertain," undermining public trust. Simultaneously, well-intentioned advocates might cherry-pick the most alarmist AI predictions to bolster calls for radical action, which, if proven overzealous, can lead to disillusionment and skepticism.

Transparency is as important as accuracy, in this case. It is not a single AI dictating outcomes, but a system where multiple models offer risk assessments, with clearly stated probability levels alongside transparent explanations of potential consequences. Additionally, people are responsible for their own part in the decision-making process: policymakers should understand how models function, avoiding bias towards the technology. Open, data-driven discussion around uncertainty is useful against the manipulation of the facts. AI might offer a better description of possible outcomes or patterns, but people must be cautious about their own biases when using that information.

Patents and Generative AI in Materials Science

Finding materials that aid our environmental agenda (such as "green" batteries, bioplastics that effectively degrade, and efficient solar cells) is a search for where generative AI could be useful. The challenge is not just finding groundbreaking concepts but addressing existing systemic constraints and potential new problems in our economic system.

Let's discuss the current manifestation of one of these problems. AI models are often trained on large datasets, including publicly funded research datasets that might have originated from underfunded university labs. When this information is used for a private AI capable of creating new material formulas, how should a patent system built for individual inventor rewards be retooled to address this issue of ownership? This is not merely a theoretical concern; we can examine ongoing court cases where such disputes already hinder the deployment of potentially sustainable tech.

A major concern that is growing as AI gets better at "inventing" is that potential profits concentrate even further within large technology firms possessing significant computational power and datasets others lack access to. The result is a troubling potential future: AI for "eco-innovation" inadvertently widens the divide between wealthy corporations and researchers struggling to secure traditional grants. Progress could be undermined by legal battles over who owns the AI's inventions.

Mitigating these risks while ensuring innovation is rewarded needs a multifaceted approach. It could include open-access clauses attached to publicly funded research datasets, a field where policy needs to adapt rapidly to the pace of AI growth. New systems to value data contribution (beyond financial backing) will also become vital. Perhaps a portion of any profit or royalties from groundbreaking materials "invented" with publicly funded datasets gets funneled toward basic research grants, creating a self-sustaining cycle. Furthermore, we might need to move beyond our focus on "invention" in patent law. Recognizing AI models as creative collaborators instead of mere tools requires significant philosophical and legal shifts, where open collaboration with these systems results in shared gains, not competition.

AI as an "Earth Systems Optimizer"

Even advanced AI climate models or resource simulators focus on specific segments of the environment, such as ice caps, ocean acidity, and forest fire risks. In the future, AI might become a vast knowledge network spanning our entire planetary Web, capable of modeling connections people currently do not perceive. Here are some potential benefits of a large AI system:

- *Cross-domain predictions:* It may be possible to create an AI-powered model that does not just focus on a single biome but that connects many types of information. For example, the system may use rainforest information, and that of distant deep-sea currents, as well as the insect population fluctuation impacting agriculture thousands of miles away, leading to an evaluation of potential human displacement and urban resource stress. Suddenly, environmental impacts can be revealed through a map of potential chain reactions, which may offer more time to plan adaptive responses.
- *Planetary safeguards:* This interconnected approach to AI environmental modeling involves the idea of planetary boundaries being monitored in real time. It can help scientists obtain a better understanding of how much disruption and stress the Earth's biosphere can manage before large-scale collapses occur. These AI-powered models may alert us not just to specific disasters but provide crucial warning years before ecological systems are pushed to the breaking point.
- *Hyper-personalized impact awareness:* The "Earth Systems Optimizer" concept is not solely for global policy. Consider an AI system with this global awareness tailored for individuals in a constructive way. The AI could provide us with possible courses of action based on where we live, our choices, and the global impact chains. The impact of buying an imported food product may be recalculated when AI factors that region's vulnerability to predicted drought. This way of using AI connects us to distant events with tangible impacts, fostering better-informed choices.

Addressing Concerns

While offering immense potential, using AI so extensively naturally involves some challenges and controversies. Let us consider some issues in this debate:

- *The "Complexity Conundrum:"* AI models as "planetary optimizers" will inevitably result in immense data needs and a level of complexity humans cannot fully grasp. It requires a new level of public education, transparent visualization tools, and explainable AI to prevent AI from being dismissed as too abstract or manipulative.

▪ *Beyond mitigation, into restoration:* These planetary simulations should not solely focus on prevention but on motivating action. This is a change in mindset, where we no longer just use AI to monitor problems to using its predictions for early actions. Can we imagine a future where AI guides eco-engineering efforts, not just on a planetary scale, but tailoring biodiverse reforestation models to the specific conditions in your town?

Using AI on a very large scale does involve some serious ethical questions. We must be careful in how we approach the topic, ensuring that every person has a voice in determining the future of such a system.

AI in Environment Case Studies

Case Study 1: The rainforest and ocean data

Problem: Conservationists focusing on protecting a critical rainforest region grapple with fragmented data. Local deforestation problems are occurring, but it is more difficult to quantify threats like the impact of changing rainfall patterns and upstream damming projects on the biome.

AI Solution: A research consortium partners with tech developers to deploy a custom AI "health tracker" spanning the entire rainforest basin and linked to data ranging from distant ocean temperature sensors to regional weather modeling. The AI model reveals unexpected factors: soot from wildfires hundreds of miles away is carried over the rainforest, altering atmospheric chemistry and stunting vital plant growth.

Impact: This insight changes the way the local biome is viewed. Negotiations with neighboring regions around pollution reduction become pivotal to saving the rainforest. This interconnected ecosystem revealed by AI creates an unlikely alliance for the sake of ecological stability.

Case Study 2: An AI-planned urban oasis

Problem: Major cities are suffering from the increased temperatures caused by climate change. They implement well-intentioned "greening" programs with mixed results. Tree planting campaigns are not coordinated, with unsuitable species chosen or planted in locations blocking valuable solar access for energy infrastructure.

AI Solution: Collaborating with urban planners, an AI model uses a combination of satellite imagery, 3D urban mock-ups, and citizen-submitted data on the presence of local wildlife/pollinators. The AI identifies not just ideal green zones but specific mixes of flora based on local needs. It guides an environmentally-sound change in the cities' approach that boosts biodiversity, creates cooling corridors, and maximizes green rooftops for urban agriculture potential.

Impact: Urban spaces are transformed, adapting to the built environment without disrupting it. Residents who first resisted the changes experience the benefits of cooler streets, cleaner air, and returning wild bird populations. These positive changes were all guided by a system balancing human and ecological well-being.

Case Study 3: Supplying aid with climate considerations

Problem: Humanitarian aid post-disaster remains reactive. A refugee camp is created after a flood displaces thousands, leading to resource depletion around it, sanitation issues, and potential secondary health crises for an already vulnerable population.

AI Solution: Relief agencies begin deploying predictive AI tools designed to model these cascading impacts. An AI model fed regional climate data, past relief camp trends, and current resources anticipate needs on a granular level, the ideal location for the camp considering access to distant water sources, as well as which medical supplies are most needed months from now.

Impact: Aid does not merely address immediate crises but becomes a tool for anticipating and mitigating secondary issues. This leads to a more sustainable distribution of resources, promotes long-term health within the displaced population, and ultimately eases the strain on overwhelmed local responders.

AI IN TRANSPORTATION

Let's imagine a future transportation scenario. You step outside, and instead of hailing a taxi or going to a bus stop, you speak a destination into the air. Moments later, a special vehicle silently arrives, its route dynamically determined by an AI network that manages the transportation needs of an entire city. The vehicle navigates through traffic, not guided by lanes but by real-time traffic flows analyzed for maximum efficiency and safety. Your journey has a broadcast going, with a news update tailored to your commute time, the day's pollution forecast shown on the vehicle's windows, and a suggestion for a detour based on AI-predicted crowds at your preferred café.

This scenario is not just a science fiction story: it is a glimpse into the world transformed by AI. Transportation, the act of moving ourselves and our goods across cities, continents, and oceans, is important to modern life. Yet how we travel is limited by congestion, delays, and a struggle against infrastructure designed for vehicles, not people. AI presents the opportunity to fundamentally redefine transportation by adapting movements to our individual and collective needs.

This utilization of AI in transportation also has some ethical challenges. How do we balance efficiency with personal privacy when an algorithm manages the transportation needs of entire cities? Does this lead to a world of easy movement or one where individuals lose autonomy over something as basic as how they access their workplace? Will AI-optimized logistics help people in underserved regions or support existing inequalities?

The answers to these issues can help create a transportation revolution that is equitable, sustainable, and aligned with human values. The future of transportation powered by AI will be determined by the choices we make. Let's consider the potential and pitfalls, exploring how AI can help us move through our world.

AI and Its Effects on Transportation

AI will have access to large amounts of transportation data that comes from roads shipping lanes, and buses or trains. All of these modes of transportation must be on schedule and adapt to the citizen's requirements in real time. This is the "New World" of transportation that can be created by AI. Some important aspects of this system are as follows.

Driverless vehicles and "smart" traffic monitoring

Driverless cars will be part of a "smart" transportation systems, as well as cameras that can help analyze road conditions for pothole prediction. Early-stage projects in several cities already demonstrate this technology. AI in city vehicles detects road wear far more reliably than human inspection, preventing damage before it occurs. Imagine a "smart" pavement embedded with sensors constantly feeding condition data to an AI.

Intelligent traffic lights can respond to real-time demands to ease congestion. Some cities test systems where AI-controlled lights actively prioritize buses and ambulances or respond to unusual surges, such as when a concert ends, and cameras send data to an AI, ensuring smoother traffic outflow.

Pedestrian flows influence urban design and AI can effectively manage them. During major events, anonymized location data analysis can suggest dynamic walking path changes or temporary barriers easing crowd flows. If we extrapolate this idea, we could potentially have an AI analyze daily commuting patterns over months, revealing underutilized sidewalks or dangerous locations. Its insights would ensure that new city plans focused on people, not just vehicles.

Logistics and shipping

AI can create efficient routes for shipping fleets, anticipating storms in distant ports; it may be able to optimize shipping routes by completely changing them. Projects exist where AI uses weather modeling, historical delay data, and port activity monitored via satellite feeds to suggest "unconventional" paths that may seem longer but result in fewer days of lost or wasted fuel. This approach challenges current industry methods.

AI can manage logistics on a global scale. It can direct the stacking of containers based on predicted departure orders, guides picking robots on ideal paths based on real-time orders, and even factors worker break schedules into loading dock assignments. Efficiency is not solely about the truck on the road, but the interconnected flow of goods in every stage in the supply chain.

Solving public transit issues

AI can direct resources to where they are needed. Start-ups and some smaller transit agencies have experimented with this idea. Instead of rigid bus routes, they have a pool of smaller vehicles that AI deploys as demand requires. This is not merely high-tech ridesharing but addressing critical needs. AI identifies a cluster of wheelchair users needing transport from a medical facility, and instantly redirects the resources available.

AI may be used for predictive maintenance, but this type of use can be expanded. A route with increased pollution can be improved using data. AI can then become a tool creating safer, cleaner cities for pedestrians and travelers.

Transportation Improved by AI

AI can enhance the routine of moving, driving in traffic, searching for parking, and riding crowded trains. AI also has the potential to fundamentally reimagine how we interact with the concept of transportation itself.

AI can influence roads, self-driving cars, robot taxis, and automated highways. It can also provide unexpected benefits:

Streets and data networks

The asphalt in streets could become part of the intelligent transportation system. Traffic cameras have now evolved from merely recording violations to gathering vast real-time data on urban mobility patterns. AI can analyze this data to dynamically adapt signage, manage flows around accidents, or send predictive alerts. It transforms existing infrastructure into a sensor network, creating a more proactive transportation system.

The global supply chain

AI can optimize shipping routes in response to weather disruptions and geopolitical events that create supply chain risks. AI can analyze goods throughout the world and predict where vital raw materials might be delayed, thus encouraging businesses and governments to take action before shortages hurt crucial industries. This type of "smart" global supply monitoring can affect people's daily lives, such as ensuring that grocery shelves are full or medicines are readily available.

Commuting

AI's ultimate achievement in transportation could be eliminating much of our travel time. AI could also offer new choices in *how* we commute. AI can manage a range of important functions, such as generating suggestions about pollution levels, giving personalized news relevant to a journey's duration, or connecting riders on public transit who have shared professional interests. AI could support human interactions or enhance the commute by helping us spend the time in a more efficient manner.

Challenges in transportation

It is vital to acknowledge that there are challenges that AI alone cannot resolve:

- *Privacy:* When AI optimizes movement patterns, what safeguards keep that data anonymous and inaccessible to corporations seeking new monetization strategies or authorities who wish to track individuals? We cannot have AI-assisted movement at the cost of surveillance disguised as convenience.
- *Exacerbating inequity:* Optimization often favors efficiency; could that come at a cost? Will underserved communities or those without smartphones find gaps widening as AI reshapes public transit and investment prioritizes "profitable" routes?
- *Trust:* Will all users readily cede control to a self-driving vehicle? Early AI projects failed because they did not understand how humans react to machine decisions. A safe ride with no human driver might still cause passenger anxiety if the AI cannot explain its reasoning for its decisions.

We must emphasize that successful AI deployment is deeply connected with social, legal, and ethical frameworks. Technology exists within a complex human world. Ignoring this fact risks the failure of the system.

AI in Transportation Case Studies

Case Study 1: Smarter streets

Problem: Urban planners often make decisions based on outdated traffic models or limited public surveys. This leads to infrastructure mismatches, such as too many lanes where congestion persists and poorly designed pedestrian crossings.

AI Solution: A city partners with a tech start-up specializing in anonymized location data analysis. Through opt-in apps and street camera sensor networks (with citizen privacy controls), AI tracks daily mobility patterns, such as how actual flows differ from the projected scenarios. This reveals the need for a pedestrian overpass near a school zone, not at a major intersection, as believed. AI also uncovers how poorly timed signals cause traffic problems, and so makes changes to the lighting schedule.

Impact: The AI implementation improves the traffic situation for people. Commuting times decrease because of the smarter flow coordination. Tax money gets deployed strategically based on proven need, not assumptions. Urban landscapes adapt based on real behaviors, becoming intrinsically safer and more human-centric.

Case Study 2: The rural commute

Problem: Public transit outside major cities is often poor, infrequent, and inconvenient, driving dependent on cars even for those poorly served by this model. Rural areas face both individual isolation and a drain on economic potential.

AI Solution: A county pilots a ride-sharing program and uses AI as part of the program. Instead of fixed routes, a dynamic allocation system factors live requests, vehicle availability, and road conditions to generate micro-routes on demand. Elderly residents gain independence in accessing medical appointments, and commuters traveling to a distant transit hub experience reliable rides based on actual train timetables, not arbitrary bus schedules.

Impact: Rural transit significantly improves. AI makes it possible what conventional fixed routes simply cannot manage cost effectively. Community connections are enhanced in these shared rides, boosting participation in events from job fairs to farmers' markets. AI helps the people in various regions improve their living conditions.

Case Study 3: Port management

Problem: Global supply chains suffer from reactive planning, port closures due to storms, geopolitical incidents, and labor unrest, leading to shortages and price instability.

AI Solution: A major port authority builds a custom AI platform fed not just traditional data but news from around the world in different languages, shipping pattern monitoring, and social media trend analysis. The AI model begins issuing "disruption risk indexes" weeks in advance, allowing importers to reroute or build additional inventory buffers preemptively. It even detects anomalies, such as increased repair ship activities in a port known for political instability, a possible signal to expect delays previously unforeseen.

Impact: Supply chains become more resilient. Companies save money through AI planning, and consumers benefit, too. AI can provide actionable information early so that the whole system has a chance to adapt.

AI AND SPACE

The exploration of the cosmos began with using star charts and telescopes, and now it is using data streams. Within the information collected by probes and space observatories lie answers to fundamental questions of our universe – if we can discern them. AI can transform complex data into actionable insights with speed and scale beyond human capacity. It can search the data on distant starlight for signs of new planets or guide spacecraft through hazardous trajectories. AI will support new space discoveries.

AI is like an important tool in the astronomer's toolkit. AI can help locate exoplanets near distant suns, navigate the hazardous orbits of asteroid fields, and monitor the Earth. We should consider the ethical implications of using AI in space exploration. If AI detects the patterns within datasets that reveal something entirely new, has it made its own type of discovery?

AI in space science will be an important tool for understanding information. There is still a need for ethical stewardship. Within algorithms analyzing starlight data may be biased. The responsibility of building trust in AI space systems is critical, especially because of autonomous space vehicles.

AI as a Tool for Exoplanet Discovery

Finding exoplanets is an immense data challenge. Telescopes generate tremendous volumes of information and detecting subtle changes in starlight (which identify distant planets) is incredibly difficult. This is a problem where AI can assist. It can easily manage large amounts of data to find new patterns humans might have failed to identify.

AI makes this process faster and more comprehensive. AI algorithms can be trained to recognize anomalies and identify new types of planetary systems. This effort could include smaller, Earth-like planets or those with unconventional orbits, reshaping our understanding of common planets in the galaxy.

Beyond identifying exoplanets, AI also allows us to analyze their atmospheres. By studying the light signatures passing through a planet's atmosphere, we can determine a planet's chemical composition and search for the components needed for life. AI may be useful in the search for extraterrestrial life.

Finding a promising Earth-like planet through AI analysis is just the first step. These discoveries still require traditional telescoping methods for deeper study and to verify initial findings. AI can accelerate the process of exoplanet detection and help us answer some of the most fundamental questions about the universe and our place within it.

AI Monitoring the Earth

Satellites orbiting Earth collect a large amount of data. They track weather patterns, changes in land use, and minute shifts in sea levels. This information could be used to improve crop yields, protect coastal areas vulnerable to rising tides, and give communities advanced warning before severe weather events. AI is the essential tool to make these benefits a reality.

While AI can reveal predictive insights from satellite image data, we can also train algorithms to detect early signs of drought, forecast harmful algal blooms, or identify deforestation patterns in protected areas. Using AI allows for dynamic analysis, offering actionable information to farmers, emergency managers, and policymakers worldwide.

There are serious questions that remain. When does "beneficial" monitoring become excessive surveillance? Satellite data combined with AI could allow near-real-time tracking of movement, resource use, and potential environmental crimes. While there are optimistic scenarios (curbing illegal fishing, for instance), misuse of this data raises profound ethical concerns about privacy and authoritarian control.

Here is the challenge and potential opportunity of using AI as a monitor for Earth: Can we use it to find important insights without hurting individual privacy rights? Developing frameworks for data access, transparent AI models, and responsible governance is as critical as deploying the technology. Done right, we can utilize AI to monitor the health of our planet better than ever before. Done without safeguards, we may have unforeseen and potentially troubling consequences.

AI and Threats from Space

Near-Earth objects, from small pieces of space debris to larger asteroids, present a silent but ever-present risk. Today's challenge is not simply knowing asteroids exist but having enough warning to prepare if one is on a collision course. Add to this the growing amount of human-made space junk orbiting our planet, a potential hazard to vital satellites and astronauts. AN AI-powered defense system is a practical necessity.

AI excels at real-time analysis of immense datasets, such as those from telescopic surveys, satellite observations, and orbital simulations. AI can use this data to rapidly identify and predict collision trajectories for high-risk objects. Faster detection offers valuable extra time for scientists and governments to calculate a plan of action.

One exciting possibility involves AI optimizing potential interception scenarios, where it may move an asteroid off-course or collect dangerous debris. It could even lead to autonomous countermeasures deployed before humans can fully assess the situation when time is critical. We must proceed with extreme caution in these types of situations.

Can we fully trust rapid, AI-guided decisions that respond to potential space threats? What kind of redundancy systems are needed in case of errors, hacking, or unexpected scenarios? An automated reaction escalating a false alarm could inadvertently spark major international confrontations or destabilize treaties, which are risks we cannot afford.

Ultimately, planetary defense is not about giving all control to an AI system but using it as a component within a larger response system. AI can constantly monitor the sky, alerting us to danger early. Yet the critical judgment decisions about acting on those alerts still require careful human consideration and global cooperation. AI, in this case, should not be acting alone but as an integrated assistant alongside scientists, strategists, and policymakers.

AI and the Data Issues in Space Exploration

We confront a unique problem in space: we can gather data far faster than we can send it back to Earth. Advanced telescopes and orbital sensors generate unprecedented volumes of information. These massive datasets hold knowledge about our planet, distant galaxies, and asteroids vital to understanding the solar system's formation. Yet, with limited downlink bandwidth, this information may not be fully utilized.

AI can analyze space data as well as efficiently manage it. AI can be thought of as a "smart" assistant. Instead of sending raw, unfiltered data, it can pre-process information onboard the object collecting the data. Prioritizing specific findings, detecting anomalies, and intelligently summarizing trends reduce the amount that needs to be transmitted while maximizing its scientific value.

AI could also help manage data compression. It can find subtle patterns within data streams, enabling smarter lossless compression and even identifying sections likely to contain noise or redundant information. Imagine transmitting twice as much scientific data using the same bandwidth. AI may be able to accomplish this type of efficiency. What if AI misidentifies the most valuable data for transmission? Perhaps those seemingly redundant images contain unexpected phenomena overlooked by the algorithm. Can we build enough flexibility into the system so scientists can change parameters? Is the AI transparent enough to understand why it selected the parameters that it did? AI can help us solve the data challenges in space exploration. In order for it to be used effectively and not accidentally discard groundbreaking information, it needs to be utilized in collaboration with scientists. Transparency and the ability to adapt AI approaches are critical to make space data management another AI success story.

AI in Space Sciences Case Studies

Case Study 1: Using space data

Problem: A large-scale international collaborative mission involves multiple orbital telescopes generating large datasets far too quickly to be fully transmitted back to Earth. Traditional approaches would create a frustrating situation: either limiting research bandwidth by focusing on subsets of data or delaying analysis as massive datasets queue up for download.

AI Solution: An onboard AI system is started by mission scientists. This AI has been trained on simulations of the potential data types the mission might gather. Equipped with this knowledge, the AI prioritizes "signals of interest" over background data. This high-value data is downlinked during communication first, accelerating the analysis speed. Further, data compression algorithms developed by AI allow larger chunks of scientific data to be squeezed into the same transmission window.

Impact: Scientists do not have to wait months or even years for the entirety of a dataset. Those initial AI-filtered results reveal unexpected phenomena, allowing adjustments to telescopes' observations in near real-time to focus on emerging areas of discovery. AI facilitates an adaptive loop for exploration instead of rigid pre-programmed observation cycles.

Case Study 2: Protecting orbits through AI efficiency

Problem: Many satellites, research, commercial, and governmental, orbit the Earth. This presents a growing problem with space debris; even tiny fragments can threaten the assets we rely on for communications, weather monitoring, and scientific research. Current collision prediction systems are reactive, and their accuracy suffers from incomplete data on orbits and the chaotic motion of debris.

AI Solution: A deep learning AI is fed years of collision prediction data, learning subtle patterns of orbital shifts and the probability of debris movement based on material types. This system is deployed globally by satellite operators. The AI acts as an efficient aggregator: Instead of raw tracking data streams, it identifies only high-risk encounters and the assets that need to take a course adjustment.

Impact: Satellites do not rely on overly cautious maneuvers each time the system triggers a warning. This reduces costly downtime and improves efficiency by refining threat prioritization with AI predictions. Most importantly, the space economy can expand confidently and sustainably, protected by a more intelligent traffic management system.

AMAZING FUTURE INDUSTRIES

If humans could fly, we would not stop at building airplanes, and we would have colonized the skies. AI is our "evolutionary leap," so the issue is not what limits us now, but how quickly we can adapt to the new technology. These new industries show what happens when AI becomes a tool for rethinking the possible. Will they bring universal benefits or create new dilemmas? Only careful stewardship will decide the ways AI can be used effectively.

1. *Asteroid resource prospecting and utilization:* Mining near-Earth objects rich in rare resources for new space development

2. *Space habitat construction and orbital manufacturing:* Transforming space into a new industrial frontier, utilizing microgravity and abundant solar energy to create products unavailable on Earth and constructing off-planet settlements

3. *Personalized organ printing and gene therapy:* On-demand organs tailored to your DNA and precision gene editing revolutionize health care, extending lifespans but raising issues of accessibility and genetic modification ethics.

4. *Quantum computation and applications:* Quantum-powered AI changes drug discovery, financial modeling, and encryption, offering unprecedented computational power and potential security threats.

5. *Brain-computer interfaces and augmentation:* Direct brain-machine communication helps those with disabilities but raises questions about technology and the self.

6. *Synthetic biology and biofabrication:* Designing microorganisms for targeted medical applications or environmental cleanup, potentially creating customized new life forms with vast and uncertain future applications

7. *AI-powered immersive experiences:* Virtual reality, haptic feedback systems, and multi-sensory simulations change our understanding of the real and digital worlds for both entertainment and therapeutic purposes.

8. *AI-driven climate adaptation and resilience:* Integrating AI for climate modeling, weather pattern prediction, and proactive infrastructure updates to combat the global impact of climate change

9. *Urban vertical farms and resource management:* Feeding the increasing number of people in cities efficiently by turning skyscrapers into farms with AI-controlled cultivation and innovative water and nutrient circulation approaches

10. *Advanced robotics and exoskeletons:* Improving human strength and movement through AI-powered robotic enhancements, revolutionizing industrial fields and offering extraordinary new options for individuals with disabilities

11. *Quantum-secured communication networks:* Quantum-enhanced technologies ensure absolute data security, enabling confidential communication across various industries while implementing privacy protocols.

12. *AI-enabled personalized health care:* Utilizing AI-powered diagnostic tools, continuous health monitoring, personalized disease treatments based on an individual's specific medical data, and proactive preventive measures

13. *New forms of sustainable energy:* Solar power stations constructed in orbit, advanced waste-to-energy solutions, and innovative renewable energy storage systems with long-term power availability

14. *Carbon capture and geoengineering:* Developing scalable carbon sequestration technologies and large-scale climate intervention technologies using AI-powered approaches to counteract global warming's destructive impacts

15. *AI-created art, music, and entertainment:* Helping with artistic expression as AI algorithms become integral tools for creativity, challenging perceptions of art, and offering unprecedented interactive experiences

CHAPTER

3

AI in *Emerging* *Markets*

INTRODUCTION

Our ideas of what AI looks like are often influenced by the ideas from technology companies in Silicon Valley. You might think AI is largely concerned with self-driving cars and "intelligent" digital assistants. To understand the transformative power of AI on a global scale, you need to consider AI's influence on emerging markets. These dynamic regions are less influenced by legacy systems and traditional approaches to business. For emerging markets, AI can become the foundation for designing more equitable, sustainable societies.

Consider how great the difference is between building an AI-guided system for optimizing traffic in a soon-to-be-built megacity versus adding that technology onto an already-existing city. How the technology is integrated can affect its ability to accomplish its goals. AI can help societies with serious issues like food insecurity, intermittent electricity, or overburdened health care systems.

Of course, this is not about developed nations providing ready-made tech solutions. Emerging markets have unique strengths, localized knowledge, a deep understanding of specific needs, and the motivation to attempt new business endeavors. AI, in this type of situation, could be used as a collaborative tool where localized datasets train algorithms to predict specific crop diseases, pollution patterns, or optimal transportation routes within their unique environments. These solutions could be adapted beyond the local area, generating valuable insights far beyond their borders.

We cannot, however, ignore the challenges. AI development demands investment in local infrastructure, training STEM workers, and building data protection safeguards. Done ethically, emerging markets become not just adopters of AI, but places for innovative approaches benefiting the global AI landscape. The solutions devised out of necessity in these regions might offer unexpected answers to some of the toughest challenges even wealthy nations have not solved.

This chapter discusses the real-world promise of AI, spotlighting innovative projects and examining where early attempts show signs of immense potential or significant failures. Without the proper infrastructure, education, and safeguards, even the most advanced AI risks furthering inequality rather than alleviating it.

AI could cause further problems in the world or it can help solve them. Understanding the dynamics in emerging markets is not just about altruism but enlightened self-interest on a global scale. Let's explore how AI revolution could reshape our world and what needs to happen to ensure its benefits are broadly shared.

AI IN EMERGING MARKETS

Emerging markets reveal how AI can be used practically. A smartphone-based crop disease early warning system in Kenya and a drone that optimizes water routes during droughts in South Africa are two of the innovations from regions with limited infrastructure. AI can be an important tool in creating novel solutions to problems, and it can be incorporated at the earliest stages of planning.

There are still challenges in implementing AI in emerging markets. Access to reliable internet, lack of skilled AI expertise, and skepticism about technology are some of these issues. Some people in wealthier nations may consider these issues and assume emerging markets are unprepared to effectively utilize AI. That is the wrong mindset. Emerging market regions are potential centers of innovation.

There must be a radical shift from current development models to tap into the potential markets for AI. The benefits of localized AI platforms built on specific datasets in performing useful actions are numerous, such as identifying crop threats unique to a region, predicting pollution patterns based on local factories, or devising efficient transport solutions within bustling urban centers. Collaboration networks and ethical data-sharing practices could allow local solutions to be adapted and applied elsewhere, benefiting global AI development from the earliest stages of a region's development.

Implementing this sort of approach requires more than altruism. Emerging markets need investment in local data centers, STEM education programs to educate AI developers, and the transparency necessary to build public trust in complex technologies. Successfully implementing AI solutions in the beginning stages of development will boost regional development and provide a fundamental change in our understanding of what AI can do. The AI innovations emerging (for necessity) from less-wealthy nations could lead to solutions impossible to engineer in the labs of the developed world.

AI Adoption in Emerging Economies

Emerging markets represent enormous potential for the development and implementation of AI applications. We must avoid pigeonholing AI as a tool for futuristic smart cities and focus instead on how it can address essential needs crucial for development.

Agriculture

In wealthy nations, AI-guided drones and automated tractors assist farmers. In a region with limited resources, a system sending text alerts directly to farmers may be far more valuable. AI analyzing satellite imagery alongside localized weather data delivers critical real-time advice: when to plant based on predicted rain patterns, which crop varieties fare best against a likely pest threat, or the most efficient use of limited fertilizer. AI can provide all of this agricultural expertise at a low cost.

- *Beyond subsistence farming*: AI can show the links between agriculture and commerce. AI-powered supply chain optimization could allow farm collectives to bypass exploitative middlemen. Predicting commodity price fluctuations could give small farmers vital knowledge and an equal negotiating opportunity.

Health care

AI can assist with health care in limited-resource settings. It would not replace physicians, but support them:

- *Remote care with AI support*: Chatbots can manage preliminary symptom queries, freeing time for severe cases in an understaffed clinic. Portable AI-powered devices can interpret ultrasound scans where a trained specialist is unavailable, speeding up diagnostic times.
- *Prevention first*: AI can provide treatment support and early detection capabilities. Predictive models built on localized data (such as that from prior outbreaks and environmental factors) help identify emerging health crises that underdeveloped health systems could otherwise miss. AI could save lives through careful preparation.

Small business

Small and medium-sized enterprises form the economic backbone of most emerging markets. However, access to loans, inventory forecasting, and even secure payment systems are often lacking. This is where AI-powered solutions designed with real-world challenges are useful:

- *The end of financial exclusion*: Traditional banks rely on credit history small businesses often lack. AI models using alternative data (such as utility bill payments, and social media activity) can provide a more accurate depiction of creditworthiness. AI would not replace bankers entirely but create new ways for businesses to obtain credit and expand.
- *Customer insights for local stores*: From inventory forecasting to understanding neighborhood consumption patterns, AI tools empower smaller operations. At the local level, family-run stores can make smarter stocking decisions.

Smart infrastructure

Efficient public services are the foundation for growth. AI should be built into the design of sustainable, equitable infrastructure:

- *Managing megacities*: Traffic, pollution, and waste management require centralized management. AI models analyzing complex urban data could suggest congestion-mitigation strategies, pinpoint illegal dumping sources, or reroute service resources with maximum efficiency without radically changing existing physical structures.
- *Rural power grids*: AI is crucial in regions far away from central power grids. Optimizing resource use of dispersed sources (such as solar, and small-scale hydroelectric) is important to keeping businesses and schools operating. Local models can then be further adapted for outer regions.

Important caveat

We cannot lose sight of the limitations of AI solutions. AI apps cannot fix a lack of fertilizer or shortage of doctors. AI tools empower communities and resource-strapped governments to use what they have more wisely.

AI Adoption Case Studies

Agriculture

- *Wefarm (Kenya)*: This farmer-to-farmer knowledge-sharing platform is simple. An SMS-based system allows farmers to ask questions and get crowdsourced answers. AI manages it: classifying queries, routing them to those with relevant expertise (even across language barriers), and identifying trends to guide NGOs in sending aid where it is most needed. It is an amazing example of a "bottom-up" AI knowledge base.
- *The Microsoft FarmBeats Initiative (India)*: This project involves low-cost sensors, localized weather modeling, and AI-driven insights. The results are impressive, such as reduced input cost and better yields. Importantly, it also focuses on data ownership, which is crucial for building trust; farmers know it is their data that is being used to feed the system.

Health care

- *Apollo Hospitals (India)*: It is important to highlight large-scale adoption alongside small programs. Apollo's use of AI for oncology has improved early-stage cancer detection rates. While Apollo hospitals are privately run, the technology could be adapted for lower-resource settings if data collection becomes democratized.
- *Livox (Brazil)*: AI-powered software aids in interpreting ultrasound scans, which is especially valuable in prenatal checks in areas with a shortage of sonographers. The focus on a widely available yet underutilized diagnostic tool (as opposed to high-tech new machines) makes this solution scalable.

Small business

- *Kopo Kopo (Kenya):* This AI-powered finance approach shows that solutions can be creative. This mobile payment platform is for small businesses historically shunned by banks. The AI uses alternative credit scoring mechanisms based on transaction history, giving even roadside vendors a chance to grow their business.
- *Numalogics (India):* AI tools often focus on customer-facing operations. Numalogics manages back-end problems specific to small businesses in developing nations such as demand forecasting and supply chain optimization using local data (such as from festivals or local events) with a simple interface. This tool is only for a small number of people but effectively helps them.

Smart infrastructure

- *ZenCity (global but adaptable):* AI analyzing social media posts may sound trivial, but ZenCity has shown cities, even under-resourced ones, can gain real-time insights. Complaints about blocked drains may signal a failing stormwater system. This AI tool could be adapted for smaller areas lacking complex monitoring sensors.
- *Eneza Education, Ghana:* This company shows how smart infrastructure can surpass a built environment. While focused on education, AI tutors delivered via SMS/basic connectivity help address critical teacher shortages. Crucially, this project constantly tracks engagement data to improve, showing how even rudimentary information can lead to iterative AI advancement.

Challenges and Opportunities

AI could become a factor in global inequality if we are not careful. The challenges in emerging markets are not merely about technology adoption but issues that address the core promises of artificial intelligence.

The data-rich and data-poor divide

AI can only work correctly with significant amounts of information. Wealthy nations can track everything from traffic patterns to consumer behavior, and so they have a decisive data advantage for the algorithms they develop. Emerging markets often lack this ability to gather such large amounts of data efficiently, which is the basic digital "fuel" necessary to build AI tools tailored to their unique needs. Worse, when the only readily available AI models are built on non-native datasets, these can introduce biases. For example, an AI-powered crop disease app may fail to identify a disease strain common in Africa if it was using image data from Europe. To solve this problem, we must consider ethical data stewardship as a part of inclusive AI development.

Human capital

Even the best ideas need people to build, adapt, and maintain AI solutions. Emerging markets lose a crucial resource when talented developers, engineers, and data scientists leave for wealthy nations in search of cutting-edge work and generous salaries. These workers could return with foreign-built AI, but different regions need to have the technology customized. They need workers who can manage and integrate the technology for local systems. Technology workers are needed in their respective home regions to help support their own local solutions. Programs specifically targeting STEM education and creating incentives for AI specialists to work in their home countries will generate these solutions. Another advantage of training technology workers for their own local regions is that it builds investment confidence. If technology companies can find strong AI talent pools in various nations, they are more likely to build research facilities and collaborative innovation spaces there, supporting long-term development.

The necessity of trust

In places where there is unreliable electricity and slow Internet connectivity, there are trust issues with the implementation of technology. A lack of basic digital infrastructure can create problems for broader AI acceptance, even before we can discuss biased algorithms or unethical data collection issues. Transparency is important to the implementation of AI. Projects designed to demystify how AI works superficially (with culturally relevant examples), combined with robust regulatory frameworks to prevent data misuse, are vital. This kind of infrastructure can support the local adoption of AI.

Addressing Global Concerns with AI

Emerging markets often suffer from the worst impacts of climate change, health crises originating elsewhere, or the pollution generated by wealthier nations. Yet, with thoughtful, ethical AI integration, this can change. Local expertise combined with AI tools can potentially turn emerging markets into critical stakeholders in finding solutions that matter to everyone.

One powerful model comes in building region-specific AI platforms focused on forecasting and prevention. Based on local climate data, a system designed to forecast crop failures in the Sahel region could be valuable for local resilience and anticipating ripple effects on global food prices. Similarly, an AI-powered pollution tracking system that uses data from smaller or unregulated factories could provide early warnings to mitigate health crises within that nation, as well as downstream regions where toxins may be carried by air and water.

However, this model cannot succeed without addressing the fundamental imbalance between wealthy and developing nations regarding AI capabilities.

Data and AI-sharing frameworks must be guided by ethical principles, preventing digital exploitation, where raw data becomes a resource extracted only to leave those who provided it no further ahead. Collaborative initiatives for creating anonymized data sets, sharing pre-trained models adaptable for various use cases, or providing resources to build localized AI expertise are a part of enlightened self-interest. We address global problems far more effectively if everyone participates in the solution process.

There is another challenge: ensuring AI itself is designed for universality. For example, suppose an AI system to detect childhood malnutrition was built on data primarily from the US and Europe. In that case, it may miss critical signs in a child from a region with significantly different diets and baseline norms. It will not just fail that child. The system will further entrench biases within the algorithms that can impact global aid decisions. Addressing the "invisible harms" AI can perpetuate requires vigilance of what data is included, how success is measured, and who can adjust AI systems once their shortcomings are detected.

Examples of AI in Sustainable Development

The transformative power of AI is not just in its technological prowess but in its real-world applications that affect people's lives and our planet's health. AI is crucial in supporting sustainability. This section of the book discusses real-life stories from around the globe, demonstrating how AI is not merely a futuristic concept but a useful tool.

From the lush green fields of India to the vast blue waters off Africa's coast, AI's applications are as varied as they are impactful. These stories illustrate a crucial point: AI's effectiveness in sustainable development is not confined to high-tech labs or affluent societies. It thrives equally in remote villages and bustling cities, addressing challenges at both the local and national levels.

Whether it is through enhancing crop yields with predictive analytics, protecting ecosystems using satellite imagery, or revolutionizing health care in remote communities, each story encapsulates the essence of AI's role in sustainable development. They show AI's potential to support economic growth and ensure environmental protection and social well-being, paving the way for a more sustainable and equitable future for all.

Doubling crop yields in India through AI-based alerts

In rural India, a small-scale farmer named Arjun faced constant challenges in predicting the right time to sow and irrigate his crops. With the introduction of an AI-driven text alert system, he received timely information about weather patterns and soil health. This simple yet powerful tool, developed by local tech innovators collaborating with agricultural experts, helped Arjun make informed decisions, doubling his crop yield. The success of this program has created similar initiatives across other farming communities in the region.

AI and deforestation in Brazil

Brazil's Amazon rainforest, a critical component of the global ecosystem, has long been threatened by illegal logging and deforestation. An AI-powered monitoring system, utilizing satellite imagery and machine learning algorithms, was introduced to detect early signs of deforestation. This system enabled quicker response times for authorities and conservationists. In one notable instance, AI alerts helped local environmental agencies thwart a large-scale illegal logging operation, showcasing the potential of AI in ecological protection.

Managing overfishing in Africa with AI

Off the coast of Senegal, overfishing by local and international vessels threatened the marine ecosystem and the livelihoods of local fishermen. An AI solution combining satellite tracking and predictive analytics was deployed to monitor fishing activities and predict zones at risk of overfishing. This system aided in sustainable fishing practices and helped local fishermen optimize their catch, balancing ecological needs with economic survival.

Enhancing water conservation efforts in Australia

A national program using AI for water conservation was initiated in Australia, which is known for its dry climate and frequent droughts. Using AI algorithms to analyze patterns in water usage and predict future demands, this program helped in efficient water management across various cities. It significantly reduced waste and improved the distribution of water resources, especially in areas prone to water scarcity.

AI-powered health care in remote villages of Southeast Asia

In remote villages in Indonesia, access to health care professionals is limited. An AI-driven telemedicine program offered diagnostic support and medical advice through a simple app. Using AI to analyze symptoms and patient history, this program provided critical health services to remote populations, significantly improving healthcare accessibility.

Smart AI solutions for urban planning in Eastern Europe

In a rapidly urbanizing city in Poland, city planners employed AI to create more sustainable and efficient urban spaces. The AI system analyzed traffic, population density, and pollution data, aiding in the development of smarter city layouts. This led to improved public transportation routes, reduced traffic congestion, and enhanced urban living conditions.

The Future of Emerging Markets

Predicting the future, especially in the context of AI and emerging markets, is an endeavor fraught with complexities and uncertainties. Yet, by analyzing current trends, possibilities, and challenges, we can predict how AI might reshape emerging markets and redefine their global standing. We must be responsible with AI, since we can use it to make positive change or exacerbate existing inequalities.

Democratizing AI

In the future, AI may no longer be exclusively used by elite institutions. Instead, it can become a tool for smaller enterprises, local researchers, and community initiatives. This "democratization" can help with local innovation. Equitable access, however, is essential to prevent increasing the disparities between the wealthy and the poor.

Looking to the future

Emerging markets possess an incredible advantage, and they have the opportunity to embed AI into their infrastructures right from the beginning. Unencumbered by the need to retrofit outdated models, these new systems can integrate smarter, more efficient solutions: energy systems that optimize themselves, urban designs that are environmentally safe, and policing models without bias.

Ethical AI

In the future, emerging markets will be the leaders in developing AI solutions. Expect vigorous discussions and pioneering policies on AI transparency, data sovereignty, and the prevention of "data colonialism."

Global talent

The West's monopoly on AI expertise will be challenged. Emerging markets, as they experience successes in AI, will become leaders in talent and innovation. There will be new collaborative AI hubs, centers of excellence with a global vision, and major technology firms will invest in these AI centers.

Sustainable solutions

Let's consider a Kenyan village where an AI platform aligns waste materials with a nearby rural manufacturer's raw material needs. Local economies flourish and pollution abates. This success story helps other regions with their own sustainable development issues.

Data as a public good

Emerging markets could discover a novel approach to using data, transforming it into a communal asset. We can envision a world where anonymized health data combats region-specific diseases, offering solutions previously overlooked by global pharmaceutical giants.

Digital colonialism

There is the possibility of *digital colonialism*, where foreign investments mine data without providing any local benefit, and AI developments occur devoid of transparency. This type of colonialism is much like historical exploitations, but in a digital form. A robust policy framework to safeguard against these types of practices is essential.

The risk of increasing inequality

The use of AI could exacerbate inequalities within emerging economies. Cities that utilize AI may prosper, while rural areas may suffer, deepening internal disparities. AI should only become a tool for bridging divides, not widening them, through inclusive policies.

The path forward

The success of AI in emerging markets is not ensured. We will create the future through choices, actions, and policies. This book aims to be more than just a collection of insights. It seeks to be a catalyst for equitable AI development. By highlighting successful models, learning from missteps, and advocating for just AI governance, we can influence the direction of this AI revolution. The future of AI in emerging markets has a strong potential for success. Let's work together to use the technology so that it uplifts, empowers, and unites people.

THE HUMAN-AI PARTNERSHIP

INTRODUCTION

Much of the concern about AI is that robots will replace humans in jobs and algorithms will make crucial decisions without human input. The most transformative potential of AI lies not in displacement, but in augmenting us. There will be a human-AI partnership, a new way of working, learning, and solving problems that transcends the capabilities of either humans or machines alone.

Instead of fearing AI, this chapter explores how this partnership is important to addressing society's most significant challenges. AI has the potential to serve as an assistant for doctors, aiding them in understanding intricate medical data for faster, more accurate diagnoses. In classrooms, AI tutors could adapt to each student's learning style, providing personalized education pathways. From optimizing complex supply chains to assisting scientists in groundbreaking discoveries, the possibilities are considerable.

Of course, this partnership is not without its complexities. To thrive, we need to reimagine workforces and adapt educational systems to cultivate skills vital for a world empowered by AI. Questions of transparency and bias in building AI algorithms must be central to any ethical deployment. The danger is not just that AI gets too powerful. We may even misunderstand how powerful the relationship between human judgment and intuition and AI capabilities can be.

This chapter pragmatically examines the present use of AI and its potential in the near future. We will examine real-world examples of AI partnerships already yielding benefits, analyze where the technology is still inadequate, and consider the fundamental shifts we need to prepare for an increasingly AI-driven world. It is not enough to build "smart" technology; we need to empower "smart" users to leverage AI responsibly for the betterment of all.

THE HUMAN-AI PARTNERSHIP

Let's consider a possible future AI-powered tool. It could have the ability to redefine our thought processes, solve complex problems, and change our existence. This tool would be superior to normal physical tools, like hammers and wrenches, and be more sophisticated than modern machinery. This would be an incredible piece of equipment.

AI is no longer an idea from science fiction stories. It has become part of our daily existence. It subtly influences our choices, from the movies we watch to the products we purchase and plays a critical role in how medical diagnoses are made or how urban traffic is managed. We must address an important issue. Can we successfully use this new AI tool as a collaborative "partner" that can augment our capabilities?

In this discussion, we will consider the potent and transformative relationship of a human-AI partnership. This alliance can be likened to humanity's first encounter with fire: a momentous discovery laden with boundless potential yet fraught with significant risks. Our challenge lies in mastering AI's power, molding it to amplify our strengths and virtues while avoiding our shortcomings.

We will discuss various problems with the implementation of AI, exploring the risks it poses in the real world. This exploration is not rooted in fear, but in a profound respect for the impact of this technology and the necessity of building robust safeguards. A partnership with AI demands trust, which must be painstakingly constructed and nurtured.

Here, we will discuss thought-provoking dilemmas, uncover pioneering strategies for aligning AI with human needs, and envisage a future where the human-AI partnership unlocks possibilities that today seem like mere fantasies. The most promising future scenario is not one of humans in opposition to AI, but rather one where humans and AI work together, creating a world where our collective potential is realized to its fullest. This is the dawn of the "Human-AI Era," where we learn to coexist and thrive alongside the machines we create.

A Symbiotic Relationship between Humans and AI

The evolution of AI is not about humanity versus machines, but rather of a powerful partnership between the two. Imagine a world where AI's computational power interacts with human creativity, empathy, and nuanced judgment. This is the promise of a symbiotic relationship between humans and AI, where we are creators, programmers, and collaborators with the technology we design.

In a truly symbiotic bond, AI becomes an extension of our abilities, empowering us to address challenges that we are not equipped to solve alone. For example, consider a doctor whose AI assistant deciphers medical scans, revealing early signs of disease that a human might miss. Another example is that of an artist who dances using AI algorithms, creating new types of expression.

This human-AI relationship has its complexities. Fostering a mutually beneficial symbiosis with AI requires trust, transparency, and a commitment to ethical development. We must ensure AI is not merely mimicking our thinking but augmenting it in ways that elevate our values and address our shortcomings.

Let us consider how a flourishing human-AI collaboration reshapes our world. AI-powered systems have the potential to revolutionize industries from health care to transportation, motivating human innovation. To fully capture this potential, we must first address this technology's ethical, social, and economic impacts.

The future where AI and humans work together offers endless possibilities but depends upon the choices we make now. In the future, we must continuously question, refine, and build safeguards to ensure that AI is used for positive change.

The following sections discuss how the elements of reshaping work, collaboration, and value alignment contribute directly to building a healthy, mutually beneficial symbiotic relationship between humans and AI.

Changing work

By reconsidering our traditional views and proactively addressing economic transformations that will arise from AI, we can prevent resentment and the perception of a human-AI competition for livelihood. If many jobs become automated, it is critical to establish societal mechanisms that allow humans to use AI and contribute to society differently. This approach fosters psychological adaptation and keeps humans invested in the partnership, encouraging continued development and collaborative endeavors.

True collaboration

Understanding AI's strengths and weaknesses allows us to create working relationships that complement rather than conflict with human abilities. If humans understand that AI is not meant to replace their intelligence but augment it, they are more likely to embrace technology and find creative ways to leverage AI's strengths to pursue collaborative goals. This sense of teamwork creates a symbiotic bond of reliance and cooperation.

Algorithmic bias and explainability

Addressing bias and creating explainable AI fosters trust. People are unlikely to support and adopt a tool that seems discriminatory or that cannot be understood. Trust results in further engagement and it helps the desire to invest in the relationship. Without trust, humans naturally view AI with suspicion, undermining a truly symbiotic partnership.

Value alignment

Aligning AI development with human values and philosophies prevents a divergence where AI acts on principles outside our moral and ethical frameworks. If it acts with fairness and compassion, humans will have greater confidence in relinquishing some control in exchange for collaborative benefits. This form of ethical grounding prevents our AI "tool" from identifying people as a problem. The AI system must be trained to identify humans as necessary.

Using AI in health care

Diagnosis and Treatment: AI's pattern recognition can aid in earlier disease detection, often surpassing the human ability to spot subtle abnormalities in medical images. Imagine an AI-powered diagnostic assistant suggesting rare conditions and potential treatment paths the human doctor might have overlooked. Human insight and decision-making are still the most important aspects of diagnosis, but this symbiosis creates a powerful safety net.

Precision Medicine: AI can analyze a patient's medical history, genetic data, and other factors to create personalized treatment plans and drug recommendations. Humans remain in charge of ethical considerations, but AI provides a tool for unprecedented individualization and optimization of health care.

Telemedicine and Accessibility: In areas with limited medical access, AI-powered diagnostic tools and chatbots can offer first aid. This extends care, with real doctors focusing on complex cases while AI empowers patients to assess minor conditions and reduces the strain on overworked medical systems.

Using AI for creative endeavors

New Art Forms: AI-generative art tools provide unique avenues for expression. While some fear this diminishes human creativity, consider it a collaborative creative effort. The human provides the initial inspiration, the AI expands upon it, and the human refines the output.

Idea Expansion: AI can help writers by suggesting plot twists, brainstorming dialogue, or highlighting potential inconsistencies in the story. AI can provide material, prompting the writer to consider unexpected plot twists in the story and overcoming "writer's block."

Accessibility and Democratization: AI-powered music composition or video editing tools make creative expression accessible to those with limited technical skills. This allows individuals to focus on ideas rather than mastering complex software, potentially allowing people who traditionally could not compose music to release their own albums.

The influence of culture on the human-AI partnership

High-Trust vs. Low-Trust Societies: Cultures predisposed to high levels of trust and a positive outlook on technology could quickly integrate AI tools. In contrast, low-trust societies might be hesitant, and developers in those societies would have greater challenges to overcome to convince the people of the benefits of AI.

Collectivist vs. Individualistic Focus: In collectivist cultures, AI applications that benefit the community, such as environmental monitoring or resource optimization, might find greater acceptance. Individualistic cultures may see greater interest in productivity-enhancing tools and personalization capabilities offered by AI.

Perspectives on Spirituality and Sentience: How a culture debates AI potentially exhibiting signs of intelligence influences the relationship between AI and humans. Questions about whether AI "deserves rights" or can "possess a soul" can be problematic in some societies, potentially limiting the level of autonomy granted to AI systems.

A successful symbiotic relationship relies heavily on human adaptation and a nuanced understanding of AI's potential and limits. While this is not without challenges, if we focus on building robust ethical and philosophical ideas about AI along with its technological development, the outcome could be a world where both humans and AI flourish.

A positive view of the human-AI symbiosis is complex and requires vigilance, adaptability, and a deep commitment to ethical principles. As we address the ethical concerns of AI, our focus must remain on developing a relationship where both humans and AI can perform well and enhance successful innovation and societal benefits.

Augmenting Human Intelligence and Creativity

Augmenting human intelligence and creativity with AI means utilizing artificial intelligence technologies to supplement and expand our natural cognitive abilities and creative thought processes. This is not about AI replacing humans but instead providing tools and insights that we could not achieve on our own, empowering us to solve problems more effectively, think in new ways, and express ourselves with greater artistic depth and originality.

This is a topic that interests scientists and industry analysts alike. Research indicates that AI has the potential to significantly enhance human creativity, generating novel ideas and refining existing ones. Generative AI tools are at the forefront of this trend, assisting individuals across various fields in developing new creative boundaries and solving complex problems. Rather than replacing human ingenuity, AI should be seen as a "collaborator" expanding our existing creative faculties. In addition to stimulating creativity, AI democratizes access to vast knowledge bases and specialized expertise, fostering personalized learning and empowering innovation across numerous sectors. Maximizing these benefits means finding a balance between human judgment and AI-driven insights, ensuring people remain in control while capitalizing on AI's unique strengths.

The Core Elements of Augmentation

Expanding on the concept of augmenting intelligence with AI, we can delve deeper into its transformative capabilities. AI as an information-processor significantly enhances our ability to manage large amounts of data, revealing insights and patterns beyond human cognition alone. For example, AI's analytical abilities in medicine can detect early signs of diseases in scans with a precision that surpasses human analysis. AI can guide investigators toward novel insights and breakthroughs in scientific research by analyzing complex datasets.

The role of AI is equally impactful in regard to creative endeavors. Generative AI tools, providing diverse outputs like text, images, or music, can serve to provide inspiration, enhancing human creativity. A novelist, for instance, could find an AI-generated image that makes her consider an entirely new storyline or character concept.

Moreover, the capacity of AI to handle routine tasks and automate repetitive cognitive processes is crucial in freeing human intellect for higher-order thinking. It allows the AI to work on mundane tasks and lets humans focus on strategic problem-solving, innovation, and addressing complex challenges that AI has yet to master. This leaves more time for human creativity and strategic thinking, fostering an environment where humans can focus on nuanced, intricate problems requiring emotional intelligence, ethical considerations, and creative thinking, areas where AI still lags behind people.

The Benefits of the Human-AI Partnership

Our world has vast amounts of information and we can never process it all. AI is a tool to help us understand all of this information and assist us with managing its complexity. Consider its application in financial market analysis. AI-powered systems can synthesize real-time news, data feeds, and historical trends to suggest market patterns far quicker than any human trader. Moreover, humans are prone to habits and biases; previous experiences or preconceived notions often shape our thinking. AI trained on carefully curated datasets offers a counterbalance, proposing insights we may not have considered due to these biases. AI's data-driven methods could reveal overlooked customer segments for a business or reveal new drug targets hidden in clinical research. Finally, personalized knowledge access facilitated by AI has democratizing potential. Learning platforms tailored to individuals and AI-powered upskilling tools can provide unprecedented opportunities for people from all backgrounds to acquire new skills, assisting both individual and societal progress.

Implementing AI Augmentation Responsibly

Responsible implementation of AI involves human control of the AI-powered tools. The most effective augmentation model ensures an interplay where the AI suggests, explores, and informs. At the same time, humans can override AI outputs, especially in scenarios with ethical considerations. This is true even when applying AI to creative pursuits; the final expression and meaning always remain with the human artist. Similarly, understanding why an AI system arrived at a particular conclusion is paramount to trusting its role. Explainable AI solutions that illuminate the reasoning of AI help detect flaws, errors, and biases, creating a necessary safeguard. This also builds confidence in those partnering with AI. Bias is a pervasive danger. Feeding AI algorithms biased data only replicates and amplifies existing human problems. Proactively seeking out potential biases and designing methods to address them early in the development cycle is crucial to preventing detrimental outcomes. Finally, as AI augmentation transforms the workplace, society must invest heavily in continuous education and workforce adaptability. We cannot just focus on acquiring the latest technical skills; the ability to critically collaborate with AI will be an invaluable future workforce advantage.

Rather than fearing AI as a replacement for human intelligence and ingenuity, a healthier perception is to view it as a technological evolution akin to gaining "intellectual superpowers." AI can help our thinking, problem-solving capabilities, and artistic expression by strategically playing to our strengths while we improve its capabilities.

Human-Centric Design (HCD)

Let us consider a future scenario: Imagine a world where the tools we use do not just perform some kind of work, but also understand us. Every app, device, and service feel perfectly aligned with our needs. This scenario shows the promise of human-centric design (HCD). HCD makes us change how we think about AI, moving away from forcing humans to adapt to technology and instead adapting the technology to the humans.

In order for us to design devices for people, we cannot just make the devices "user-friendly." HCD means that AI-powered solutions will have an empathy for people used in the approach. Part of HCD involves understanding the people you are designing for, considering their frustrations and goals, and determining how a solution will fit into their lives. This design method does not start with pre-built answers but with a willingness to collaborate, test ideas early and often, and never dismiss user feedback as inconvenient.

This approach to HCD supports innovation. When technology aligns with our genuine needs, we gain powerful, useful tools. HCD can result in better business outcomes, more fulfilling human-computer interactions, and solutions that avoid becoming just another problem to manage. Consider the apps in your own life that seem designed to waste your time or make it challenging for you to accomplish your goals. These problematic apps show the consequences of ignoring human needs in technology design.

HCD is important in the development of AI systems, because they can be ethically hazardous when designers solely focus on enhancing their capability without understanding the complexities of human use. Designers must build AI systems with the right safeguards, anticipating potential biases, and prioritizing transparency. For HCD, we look beyond the technology's potential on its own and consider its potential for human use.

Technology is affected by rapid innovation, and HCD emerges as a critical design paradigm to ensure solutions prioritize human needs, capabilities, and experiences. Traditionally, technology was built on a foundation of feasibility and efficiency, often leading to systems that are complex, difficult to use, and ultimately fail to address the needs of end users. HCD fundamentally alters this approach by embedding user perspectives within the iterative design process from inception to final deployment.

Human-Centric Design Principles

- *Empathy-Driven Inquiry:* Building a deep understanding of users through various qualitative and quantitative methodologies (such as interviews, contextual observations, and surveys) is crucial to uncover nuanced needs, pain points, and behavioral patterns.
- *Prototyping and Iterative Design:* HCD prioritizes rapid prototyping, allowing users to interact with low-fidelity solutions early on. This continuous feedback loop allows for ongoing refinement and ensures the final design meets theoretical user requirements and delivers usability.
- *Co-Creation:* Stakeholders and end users should be active participants in the design process. Diverse insights and expertise fuel creative collaboration and build a sense of ownership, increasing chances of adoption and lasting value creation.
- *Focus on Outcomes:* HCD extends beyond functionality to the intended positive impacts the solution will bring. In addition to usability metrics, consideration of broader benefits and potential unintended consequences guides responsible, user-centric decision-making.

HCD offers other benefits beside user-friendliness:

- *Reduced Mismatch:* Minimizing the gap between the mental models of users and developers prevents costly development of technologies people struggle to integrate into their lives or workflows.
- *Enhanced Innovation:* By addressing real-world problems directly, HCD opens new possibilities for solution development otherwise obscured by inward-focused technical limitations.

▨ *Competitive Advantage:* Products and services designed with human-centricity deliver superior experiences, boosting user adoption and customer loyalty.

The imperative of HCD becomes even more profound in AI development. The inherent complexity and often abstract nature of AI systems call for an emphasis on explainability, bias mitigation, and ensuring solutions work harmoniously with human decision-making and judgment while maximizing the unique strengths of AI. HCD practices enable developers of AI systems to balance innovation with a strong emphasis on ethical considerations and societal impact.

The HCD Process

While real-world implementations are adaptive, a general process involves the following steps:

1.   Immersion: Becoming Informed About Users

This stage is like taking a journey into the world of your users. The goal is to eliminate assumptions and truly understand the people your design is meant to help. How can this happen?

▨ *Interviews:* Have meaningful conversations with potential users, asking open-ended questions about their experiences, challenges, and desires. It is not about confirming an idea but seeking honest insights.

▨ *Observations:* Watch how people interact with their environment and existing tools similar to what you aim to build. Notice workarounds, frustrations, and moments of unexpected ingenuity. These moments can inspire better design.

▨ *Research:* Gather additional data, such as market research, social trends, or anything relevant to your users' context. This information can help develop your design.

Immersion supports the following stages. Without this depth of understanding, your design may fail to accomplish its goals.

2.   Ideation: Thinking Creatively

With a well-rounded understanding of users, it is time to start generating potential solutions. Here is where diversity of thought is important:

▨ *Collaborative Brainstorming:* Do not just rely on the technical team. Ask developers, designers, potential users, and stakeholders with varying expertise. Differing perspectives can lead to unexpected successes.

▨ *Quantity Over Initial Quality:* Aim for numerous ideas, from practical to far-fetched. Do not pass judgement on any of the ideas immediately. The best solutions often emerge from initially wild possibilities.

- *Beyond the "Solution:"* Consider technological fixes and the entire user experience. Can a simple process change be as helpful as a new app? Explore all possibilities.

This stage sets the groundwork for refining ideas by focusing on what aligns best with the needs uncovered during immersion.

3. Prototyping: Quickly Creating a Working Model

Prototyping means giving ideas a form. The emphasis is on quickly getting something into users' hands. This approach involves the following:

- *Low-Fidelity:* Prototypes can be wireframes, sketches, or basic physical models. Quick and inexpensive creation encourages flexibility and adaptation.
- *Fail Early, Fail Cheap:* The main point of prototyping is to discover flaws early. Creating an elegant prototype that nobody can use is counterproductive. Prototype to find your design's strengths and weaknesses before spending much money and time on its full development.
- *Gather Varied Feedback:* Do not just ask, "Do you like it?" Explain how the prototype might solve problems, where it may fail, or consider surprising user interpretations your team never considered.

Prototyping exposes whether an idea is applicable in the real world or requires adjustment with minimal financial loss.

4. Testing and Iteration: Design Evolution Driven by Feedback

Think of this phase as your solution maturing. Here is how users shape this transformation:

- *Observe and Document:* Watch users interact with prototypes in as genuine a context as possible. Note every problem, moment of confusion, and positive engagement because those types of information guide you.
- *Non-Leading Questions:* Ask genuine questions to uncover true problems users face: "What are you trying to achieve here?" instead of "Can you find the X button?"
- *Open to Radical Change:* Feedback might result in minor tweaks or a complete change in design. Do not be afraid to discard your original idea if testing indicates a new direction with a better chance of fulfilling user needs.

Design should feel like a conversation guided by how your solution helps real people.

Important: This is a simplified cycle! Actual HCD involves continuous looping, refining, and retesting to create truly robust, user-centered solutions.

Why HCD Matters

Human-Centered Design (HCD), particularly in the context of AI, is essential for several reasons, each underscoring the critical intersection of technology and human needs.

Ensuring adoption and usability

AI systems designed without considering the end-user can result in technologies misaligned with user needs, leading to low adoption rates. HCD focuses on creating solutions that are not only technologically advanced, but also align with users' preferences, needs, and daily routines.

A user-friendly design encourages widespread adoption and integration into everyday life, making the technology more effective and valuable.

Addressing and preventing bias

AI systems are prone to inherit and perpetuate biases in their training data. Suppose the data lacks diversity or contains inherent biases. In that case, the AI can inadvertently become a vehicle for these biases, affecting decision-making in critical areas like recruitment, law enforcement, and healthcare.

HCD incorporates numerous types of human perspectives in the development cycle, helping to identify and correct potential biases in AI systems. This approach fosters the creation of more equitable and fair AI technologies.

Mitigating unintended consequences

AI technologies are incredibly complex systems that can have unintended and unforeseen consequences when deployed in real-world scenarios. Without careful consideration of how these systems interact with various aspects of human life, they can lead to adverse outcomes, some of which may be harmful.

HCD involves thorough consideration of an AI system's broader impact, including potential risks and negative outcomes. This foresight helps in designing safe and beneficial systems in their intended use.

Building and maintaining trust

Trust is a key factor in the acceptance and success of AI technologies. A lack of understanding of how AI systems make decisions can lead to user mistrust and apprehension.

HCD principles advocate for transparency and explainability in AI systems. By making AI decisions more understandable and relatable to users, HCD helps build trust. This is particularly important in areas where AI aids in critical decision-making.

HCD is not just a design philosophy but a necessity for AI. It ensures that AI technologies are developed with a deep understanding of human values, needs, and behaviors, leading to more effective, equitable, and trusted AI solutions.

The Human-AI Advantage

The potential use of artificial intelligence is not about replacing human capabilities but forging a dynamic alliance where our individual strengths work together. True innovation rests on creating intuitive interfaces, collaborative workflows, and AI systems that actively support human judgment and critical thinking rather than mindlessly automating. Human-centric design is an important part of this exciting collaborative approach. We will explore practical design strategies to ensure human oversight and control, explainable AI to foster trust, and proactive ethical frameworks to prevent AI tools from being used against us.

Designing for human-AI "symbiosis"

Instead of considering AI as a separate intelligence, we should start viewing it as a powerful tool, and all-powerful tools require thoughtful handling and design to show their full potential. *Human-AI symbiosis* is a logical way to advance this idea, where people supply creativity, judgment, and lived experiences, and AI supplies the processing power, analysis of large datasets, and identification of complex patterns. Designing systems created explicitly for this interplay is the key to unlocking transformative solutions.

Consider the example of a master craftsperson using an advanced digital lathe. The interface allows direct input as needed and constant refinement based on what is being created. Ultimately, the tool does not dictate the final form; the human creator has that control. We need more AI systems, especially complex ones, built with this collaborative interplay in mind. In a medical setting, that AI medical assistant is not simply diagnosing by rote but offering possibilities to a doctor who makes the final informed judgment call. A business analyst would not only receive a predicted sales trend but the power to ask their system "what if?" scenarios, revising assumptions based on their domain expertise.

We get true symbiosis when the machine adapts to us, not the other way around. Imagine AI assistants that pick up on subtle patterns in how we work, understand how we organize information, or identify recurring bottlenecks. Over time, such adaptive AI assistants learn to support us with greater ability, making us more efficient without a significant mental burden. It all lies in design that does not force humans to adjust to rigid AI-generated workflows but allows this reciprocal flow. This way, the end result is something greater than the sum of our individual abilities.

Emotional AI and affective computing

Emotional AI is the idea of machines recognizing and even potentially responding to human emotions. If designed with compassion and insight, its potential for good is undeniable. Conversely, failing to incorporate human-centric values in design could have far-reaching, negative consequences.

Consider what a breakthrough it could be if AI systems reliably read not just overt signs of anger or frustration but subtle cues that are often challenging even for humans to articulate. In mental health, an app analyzing microexpressions or patterns in voice intonation could lead to earlier intervention or offer additional support when patients cannot express their needs. It would create a digital safety net, but only if used responsibly and always coupled with professional human guidance, never taking the place of therapy itself. Emotional AI may also be used as a tool for those on the autism spectrum.

Equally vital is understanding where those fine lines and potential pitfalls exist. There is a clear difference between an AI recognizing emotions as feedback to personalize learning or enhance therapy and a system engaging in manipulation to increase sales. Consider social media sites already adept at pushing content to amplify feelings. With emotional AI, it could increase the number of anxiety-inducing in feeds if it senses vulnerability. Technology should aid in self-awareness, not become a new method of exploitation. The potential for abuse exists in any area focused on reading emotional states, making strong privacy controls and clear user autonomy nonnegotiable design principles.

Finally, it is critical to keep expectations realistic. This is not an area where a simple algorithm mirroring back how it thinks your face looks equals genuine "AI empathy." Real understanding of emotional nuance arises from data gathered across body language, tone of voice, cultural context, and individual variations. Overreliance on a single type of analysis, especially of superficial expressions, risks the harmful reinforcement of stereotypes. Successful affective computing hinges on building diverse datasets, employing collaborative teams that include psychologists, and recognizing the complexity of human emotional responses. Building these safeguards against flawed interpretations is not about slowing innovation but ensuring that a tool intended for well-being does not inadvertently turn harmful due to rushed development.

HCD in the metaverse and extended reality (XR)

The metaverse and extended reality (XR) force us to rethink the most basic tenets of interface design. They make us consider how to move away from our old ways of thinking about reality. Human-centric design can prevent these virtual experiences from feeling complicated or overwhelming to our senses. It guides us in understanding how a truly immersive environment needs to work in harmony with our bodies and natural sense of the world.

Consider designing to prevent disorientation in an AR shopping experience where virtual items overlay existing displays. If transitions happen jarringly, users will not engage; they will simply want the unpleasantness to stop. Cues on how to engage with immersive elements without "breaking" the experience and allowing people to feel safe are core HCD concerns. In addition, while advanced haptic suits and elaborate controllers may seem amusing, HCD advocates prioritizing intuitiveness. A virtual experience should not feel challenging; it needs first to leverage how we move and make sense of the world with existing abilities.

Consider also how existing spaces can be used with XR. For example, AR markers can be hidden in a historical site, prompting location-specific audio narrations and historical reconstructions overlaid at appropriate angles and perspectives. It is about building an XR "layer" that enhances what is there, not making the real world an afterthought.

Beyond how we view the world and move around, XR interfaces can also utilize our physical responses as input, whether detailed body motion tracking or even biometric data. Designing environments that subtly adapt based on a detected heartbeat spike or shift in muscle tension allows for interactive narratives or feedback. However, HCD dictates proactive ethical consideration around consent and personal data control throughout every part of development. We cannot wait until problems arise to consider solutions for biodata issues; these need to be part of the conversation from the beginning. Physical inclusivity needs the same level of proactive planning. Designing embodied XR with a limited view of physical ability creates exclusion. If an immersive dance game only uses complex lower-body gestures, how does a participant with restricted physical mobility participate? This highlights an HCD priority of building experiences with adaptable settings and various control methods.

Designing for equity and accessibility

Designing for equity and accessibility is critical for responsible AI development. When we prioritize building inclusive technology, we do not just use a checklist of items that make us "good." We need to understand new perspectives, engage in collaborative strengths with others, and develop solutions with immense potential to reach communities currently underserved by innovation.

One focus area is building inclusive AI. When AI learns through biased datasets devoid of diverse representation, it inevitably fails to work as reliably when confronted with different kinds of users. For example, facial recognition may have trouble with a range of skin tones because they were not included in the training data, or an AI chatbot might misinterpret the words for speakers outside a narrow regional dialect. Responsible HCD incorporates an understanding of bias from the beginning and addresses it before any design starts. AI development teams should include people representing various races, genders, socioeconomic backgrounds, abilities, and experiences from the initial stage of a new AI venture. With everyone having a stake in recognizing biases early on, flaws invisible to any specific set of developers can be identified. This principle extends to testing: actively seek out potential users outside the expected "average" user who can point out limitations in the system for people they share lived similarities with. Sometimes, even incentivizing this process (offering "bias bounties") can get more people involved and allows communities to contribute insights the technical team may have missed.

It is crucial to extend design inclusion beyond physical disabilities. Too often, the technology relies on users understanding complex jargon, settings, and a certain tech fluency level to leverage what a supposedly powerful AI assistant offers. Prioritizing plain-language interfaces, user-friendly onboarding tailored to varied experience levels, and tools with adjustable difficulty make an enormous difference. These efforts are important so that the AI tool can be used by many people. AI tools confined to only serving specialists will not reach their full potential in shaping society. It also extends to economic access. If only those with high-end equipment can run the AI software, even if designed without user knowledge in mind, the benefit only goes to the privileged few. Cloud-based AI options or models are adaptable to low-specification hardware and can help democratize the tech.

AI literacy is an important component of AI development. We must ensure this technology does not just feel like abstract magic wielded by the few. It cannot afford to be something misunderstood and either demonized or passively accepted just because the average person feels the inner workings are beyond their understanding. Engaging everyone takes effort. Museum exhibits designed for hands-on interactive AI experiences, simplified yet non-condescending explanations of standard AI tools for news outlets, and community-level collaboration projects can help people understand AI systems. When the average person uses AI to create positive impact within their communities, understanding grows as organically as it did when the Internet spread beyond early adopters.

Responsible AI Development

Responsible AI development is at the intersection of groundbreaking innovation and ethical governance. The excitement around AI's rapid advancement is understandable, but this enthusiasm must be tempered with a thoughtful approach. True transformation through AI is not simply about computational efficiency or swiftly solving complex problems. Its deeper impact lies in creating systems that embody human values, minimize harm, and enhance the world in meaningful ways, transcending mere efficiency.

The dual nature of AI, where the risk of harmful outcomes shadows its immense potential, cannot be ignored. AI's capabilities are like a double-edged sword: as much as they promise positive change, they equally pose risks if developed without ethical foresight. The development process must consistently evaluate the ethical implications. We must ensure a disciplined approach to predict unintended consequences, acknowledge vulnerabilities, and integrate safeguards without sacrificing innovation.

Consider the example of facial recognition technologies. While designed to assist law enforcement, biases in these systems can lead to unjust profiling, underscoring the need for more than just good intentions in development. AI systems that inherit biases from human-generated datasets do not naturally avoid these issues; they can perpetuate them. It is imperative to continuously evaluate the role of AI not only as a problem-solver but also as a potential contributor to existing problems.

In crafting AI, diversity in the development team is essential. Teams of coders alone cannot address the multifaceted ethical considerations AI demands. Involving experts in data ethics, sociology, and history and proactively seeking diverse perspectives, especially from communities that the technology could impact, is critical. This approach is not a formality, but a necessary step to identify and mitigate the team's biases before they become entrenched in the technology.

Building trust in AI systems requires robust testing in varied scenarios, external reviews focusing on bias detection, and a culture that encourages identifying problematic aspects early in development. Transparency is important, especially for AI solutions addressing societal challenges. It is about enabling the public to make informed decisions and hold developers accountable. This transparency includes openly sharing how these systems work, their limitations, and the logic behind their choices.

Finally, responsible AI development acknowledges the impossibility of "zero-risk." It is about managing the complexities and potential risks while striving for ethical and beneficial outcomes. It involves carefully considering difficult decisions and understanding that not all innovative ideas are viable or ethically sound and that some risks are inherent in pursuing advancement. The development of ethical AI is not just about technological achievement; it is about shaping a future where AI enhances human life in a way that is responsible, ethical, and aligned with our collective values and aspirations.

5

AI GOVERNANCE AND POLICY

INTRODUCTION

Building ever-more intelligent AI systems is a global endeavor that takes time and commitment. While there is immense potential to transform industries and improve lives, focusing solely on building powerful AI systems is problematic. AI governance will determine whether AI ultimately creates a more prosperous and equitable world or simply magnifies current problems at an unprecedented scale.

It is tempting to view AI governance as a "battle" between large technology companies, governments or global consensus, and local needs. This view is inappropriate. In this chapter, we will consider how governments, international bodies, watchdogs, and tech companies can make mistakes when shaping how AI functions in society. We will also discuss examples of proactive policies and public-private collaboration that, while occasionally imperfect, offer valuable lessons on responsible AI development.

From courtrooms where intellectual property law struggles to keep up with AI-generated content to cities piloting the responsible use of police surveillance technology, these real-world battles are revealing the challenges of addressing AI. This chapter highlights how legal concepts like bias, liability, and accountability can be challenging to apply to algorithms that constantly learn and evolve. The issue is not whether AI could improve healthcare, make hiring fairer, or help nations address complex challenges. Here, we will consider what kind of governance makes progress possible without worsening problems.

Our mission is to delve into the heart of important ideas, moving beyond a mere recitation of rules and regulations. The goal is to reveal the following:

- *Consequences:* When does "moving fast" with AI deployments cause problems, with real-life examples that show the costs of weak governance?

▪ *Success stories:* We will discuss often overlooked instances of governments effectively utilizing ethical AI ecosystems.
▪ *Ongoing debates:* Regulations lag behind innovation or loopholes remain even in robust systems. This is the reason why governance is never finished.

Finally, we will consider AI governance on a global scale. Emerging models in nations with differing resources and priorities may offer valuable lessons to developed and developing economies. We need global governance in this interconnected age, but success is connected to finding adaptable models, not a using single rulebook imposed upon vastly diverse nations.

AI GOVERNANCE AND POLICY

People are rapidly developing powerful AI technologies, but successful management of AI is not about who has the most intelligent algorithms. We must identify how to harness AI's transformative potential without creating new problems faster than we can solve them. AI governance is critical. It can help us decide what we can build, what we should build, and how we will manage the consequences.

Let us consider some approaches in use. For example, the new AI Act addresses issues of transparency and bias in the EU, and it is a bold attempt to guide innovation within ethical guidelines. Some nations are prioritizing rapid AI development while social safeguards are not being considered, which is a mistake as these systems become deeply embedded in everyday life. We need honest discussions about these divergent paths and how they could impact all of us, far beyond their place of origin.

International organizations like the UN seek common ground amid all these differing policies. However, finding global consensus becomes more challenging when you compare a resource-limited nation managing AI risks with basic service delivery to the concerns of a Silicon Valley company that fears overregulation will limit its creativity. Everyone feels a sense of urgency, but that urgency cannot justify hastily developed policy, especially not when those worst affected in a potentially negative scenario may not have enough power to affect its own future.

AI affects how legal issues are managed. Intellectual property cases are just one legal matter to consider. What level of liability falls on creators if AI generates harmful content? Can privacy lawsuits keep pace with the speed at which AI analyzes massive amounts of personal data? Legal issues with AI will have real-world ramifications for innovation and individual rights.

Fortunately, examples of good governance do exist. Let's examine models emerging from Europe, as well as Singapore's emphasis on public sector responsibility with AI and a Latin American development bank funding programs addressing AI bias. These cases suggest good AI governance can exist in places not typically seen as technology leaders.

We should also consider global successes, mistakes, and difficult questions AI policy raises. An AI revolution relies on fair regulations, strong institutions, and an understanding that building good AI often means first acknowledging its potential to cause harm and deciding, as a global community, what levels of risk we are willing to accept for progress.

Role of Governments in AI

As AI develops, governments must be involved in the design and application of the systems. National AI strategies are not simply about tech investments; they reveal deeply held beliefs about progress, social impact, and a nation's envisioned place in the future. By examining resources like the OECD's AI Policy Observatory, we find fascinating insights about AI management.

Some nations are using significant amounts of resources for AI research and development to outpace their rivals. However, China's ambitious AI plans and ethical challenges have resulted in some nations taking a different approach. The EU, for instance, with its AI Act and human rights-centered approach, signals that ethical considerations cannot be an afterthought, even if it slows initial implementation. This pushback against rapidly developing AI without considering its human impact raises significant questions. Is strict early regulation worth potentially falling behind in computational power? How do you foster AI progress alongside public distrust of complex, evolving technologies?

The specifics of these strategies matter. Let's analyze real-world regulations around data protection, algorithmic transparency, and liability when AI makes mistakes. Some questions we should consider are as follows:

- Is consent enough when AI constantly learns, even from anonymized data?
- How much "explainability" should creators be mandated to provide?
- Will traditional product liability frameworks still be useful under the complexity of self-learning systems?

These questions are part of an important debate because the answers, once written into law, define whether small businesses confidently innovate alongside the tech giants.

Importantly, governance is not just about what happens at the national level. Regional initiatives often serve as experimental grounds. For example, a Canadian city might pioneer novel procurement rules, making government contracts favorable to start-ups building explainable AI or a German state might set tough standards for police use of facial recognition systems. These bottom-up approaches hold enormous potential to motivate nationwide shifts in governance and result in broad global change.

Of course, the government's role is not limited to external regulation. How it utilizes AI within its own functions sets an example for others. Procurement policies prioritizing tools built with data diversity in mind can cause an entire industry to implement better practices. Missteps, such as flawed risk assessment algorithms leading to wrongful reductions to citizens' services, act as painful yet powerful lessons in why public sector deployment is important.

Finally, governance requires more than rules; it also requires skilled people to implement them. We will consider the nascent AI bureaucracy that is taking shape globally, such as countries creating dedicated ethics committees within ministries, blending internal expertise with external advisers from academia and even activist groups. This is a delicate and challenging balance of ideas. If the technology industry becomes too dominant in government, we risk *regulatory capture*, where rules favor those who should be regulated. A similar problem can occur when the people crafting AI rules do not understand how the technology works, and so the regulations they develop are ineffective.

Regulatory Organizations and International Groups

AI systems do not have international borders or jurisdictions. Governments should not act alone to fully determine how AI systems affect people. Uncoordinated national rules would stifle beneficial innovations and allow for the potential of harm. Hence, organizations that operate on a multinational and global scale are needed.

The EU, which wants to become a leader in AI ethics, the OECD's attempts at guiding ethical development, and UNESCO's focus on the social impact of emergent technologies are all important in determining the ethical constraints for AI systems. We should not only allow these large groups to determine how AI will affect people. Understanding why those organizations are powerful within their sphere and where their influence stops is crucial. Guidelines for transparency in AI may sound useful, but what real authority do these groups have to enforce compliance for private technology firms operating on multiple continents? We will explore the potential and constraints of various organizations to better understand their impact on the development of AI regulations.

There are a number of other groups involved in addressing the issues of global AI governance. The Gulf Cooperation Council brings together nations rich in resources seeking to leapfrog into AI dominance; their collaborative plans give insight into how economic priorities shape the ethics conversation in ways distinct from the West. Latin American development banks have been investing in building ethical AI expertise from the start. These groups are finding solutions uniquely suited to their challenges. The African Union is managing continent-wide initiatives for AI with limited funding in order to help those most acutely affected by potential technology harm. These emergent powerhouses bring vital perspectives and, crucially, could offer practical models other developing nations might feel better aligned with than those dictated by technology giants in more developed nations.

Governments and international bodies must work together with non-governmental organizations. We often rely on non-governmental organizations (NGOs) to expose technology abuses: a leaked internal document highlighting an algorithm perpetuating biases is the kind of scandal NGOs can reveal. Some NGOs collaborate with other industry members to create common standards. This type of partnership can streamline compliance and ease public distrust of AI systems. What happens when there are deep disagreements between technology companies and NGOs? These problems, sometimes even within civil society groups, illustrate the struggle to define what we mean by "responsible AI."

Legal and Ethical Frameworks

AI ethics are complex, but we cannot let terms like "bias mitigation" or "algorithmic accountability" become abstractions. These concepts have significant, sometimes devastating, consequences in the real world, so examining them through court cases and legal controversies can expose hidden challenges and highlight how even well-meaning developers of AI can make mistakes. Just because a program *can be* coded does not mean it *should be*.

Bias, however, can easily appear in technology designed to help. AI used in the criminal justice system might analyze crime data to recommend areas for targeted policing. Now, if the inputs contain decades of data from biased policing practices (such as targeting low-income neighborhoods), the AI is perpetuating that injustice. Simply removing categories like "race" will not remove bias inherent in where arrests tend to be concentrated. We must acknowledge these uncomfortable truths about "objective" data and question any "neutrality" claims about AI trained on flawed data. This becomes especially worrisome when predictive risk models inform bail decisions or sentencing, because people's lives can be negatively impacted.

Another ethical challenge is about how AI and creativity interact. Imagine a scenario where an AI convincingly fakes the distinctive visual style of a deceased artist. Is the result a "deepfake" piece of entertainment? Does it diminish the original artist's legacy? What are the copyright implications if a living artist has their work so flawlessly imitated that they lose income? These types of legal cases can become confusing, revealing the mismatch between existing intellectual property law and the ever-evolving nature of machine-generated creativity. Courts could create precedents here that reverberate through every art form, determining whether humans deserve legal protection when machines can replicate their talent.

While the term "deepfake" is usually associated with video or audio, the concept applies to any AI-generated content that convincingly imitates a person's style or likeness. This raises questions about the potential for deception, exploitation, and even the erosion of trust in artistic works.

Consider the impact on healthcare: doctors utilize AI for everything from complex diagnostics to treatment recommendations. This adds a layer of complexity to potential medical malpractice suits. Suppose a diagnosis follows medical protocols and utilizes a highly regarded AI tool, but both humans and machines are wrong. Who is at fault? Did the doctor not consider the analysis thoroughly enough, or is the software manufacturer (perhaps downplaying known problems) culpable? This issue is not merely about the money involved or harm, but the patients trust in a technology they barely understand.

AI hiring processes promise streamlined hiring and more "objective" assessments of candidates. These tools, however, have shown significant racial and gender bias. There are also issues about potential employees "fit" at the company. Will these tools favor those who conform to existing company stereotypes, even if unintentional? Is the promise of unbiased AI even achievable under an economic system that thrives on inequity?

AI in policing is another contentious issue. Law enforcement agencies consider AI-powered facial recognition or threat prediction models invaluable, while civil rights groups warn of dystopian futures of "pre-crime" monitoring. When flawed systems create false arrests or identify legitimate protest activity as a danger, the debate ceases to be theoretical. What legal constraints must be built into technology with such potential to curtail fundamental freedoms? What are the guidelines the law must enforce before these systems make marginalized communities even more vulnerable?

This discussion of the complexities of AI law is not meant to discourage innovation but to show why rigorous ethical frameworks are just as important as computational power. The law may seem slow to change when compared to the speed of the development of new AI applications. These cases, some shocking and some still unresolved, show that courts do not just implement AI regulations but create their guidelines.

We do not have all the answers. Is there ever a scenario where predictive policing with even slight bias is ethically acceptable if crime drastically declines? Should AI "creators" have some responsibility for the system's outcomes, much like how parents can be held accountable for a child's behavior? What happens when an AI developed under minimal ethics laws in one nation is licensed in another with far stricter constraints? Does the AI system need to follow the rules of the new country? By actively dealing with these challenging questions, we define the future of ethical and equitable AI.

Effective (and Ineffective) AI Governance

It is easy to dismiss AI governance as bureaucratic problems slowing down the "real work" of making "smarter" machines. This part of the book aims to dispel that myth. Good governance actually creates channels for the positive impacts of AI while mitigating potential harm. The cases we will analyze are like real-world experiments revealing which policies are useful for technological growth and which ones harm the utilization of beneficial systems.

Using AI successfully

Singapore is a nation with a strong focus on efficiency. Critics of AI see Singapore's use of public sector applications as potential government overreach. Singapore's government, however, made procurement conditional upon stringent privacy requirements and create regulatory testing scenarios where public safety-related AI is piloted. Valid surveillance concerns do exist in the country. However, it offers a model for governments actively fostering a specific kind of AI ecosystem rather than letting private ventures dominate the scene unchallenged.

The EU offers another rich case study with the caveat that a continent-wide effort will yield mixed results. There are successful EU programs that invested in creating AI ecosystems. Were regulations tied to funding schemes that helped small start-ups navigate data compliance in specific industries? Perhaps a specific focus on local bias audits led to greater public acceptance of AI tools built closer to home. The EU's experiments with AI systems show that AI governance should not just be about punishment, but positive incentives for developers to do things the right way.

Failures as warning signs

AI governance can be poorly implemented. Poor implementation is not just about "evil" companies using AI for harm. Some AI implementations may have started with well-intentioned rollouts of AI for tasks like benefit allocation or targeted education programs. When systemic issues, such as historical discrimination, are not fully considered, these AI models exacerbate existing problems rather than resolve them. Bias that is inadvertently trained into an AI tool deployed in a public system may take years to discover and correct. We do not just want to document failure, but the specific errors that contributed to it, such as flawed assumptions, rushed timelines, and lack of critical stakeholders in the process. The negative outcomes of poor implementation expose how weak governance creates lasting problems.

AI implementation failure can also come from public-private partnerships. It is easy to assume technology companies have unique expertise worth utilizing. However, opaque relationships, with companies gaining access to data or undue influence over standards development, create the perception of unfairness (even when no provable illegality occurs). In these instances, public trust falters. Even a successful initiative "feels tainted" when items such as procurement documents are not open to scrutiny. It can be challenging to understand the differences between collaboration and collusion in these types of situations.

Additional issues with regulation

There are instances where the law was absent, too vague, or too slow to respond to a sudden, potentially destructive use of AI. In some of these instances, NGOs have revealed wrongdoing through leaks or grassroots campaigns raised public awareness to force the government to create new AI rules. From holding AI-powered hiring software accountable for disparate results to forcing social media giants to rethink targeting algorithms, activist efforts can affect official governance. This intervention from NGOs might seem less efficient than government-created rules. However, it serves as a crucial reminder that governance, to be truly effective, is an ongoing conversation between the government and the governed.

Global AI Governance

The rise of artificial intelligence is a thrilling yet intricate challenge for global governance. It is much like the early days of computing, when the personal computer transformed entire industries. While a clear sense of shared rules and frameworks was nonexistent then, governments must act with speed and unity in managing AI.

Currently, AI governance regulation does not follow universal standards. Nations have implemented various strategies, from ambitious national AI roadmaps to tentative early forays into regulation. This irregular approach reflects the inherent complexities in balancing rapid technological innovation with social, ethical, and economic considerations.

The lack of a globally harmonized approach to AI carries several risks. Inconsistent or underdeveloped regulatory frameworks can stifle innovation in those spaces where AI promises profound breakthroughs. Conversely, a "race to the bottom" might arise, where nations with lax rules attempt to lure AI development at the expense of safeguarding against harms like algorithmic bias, misuse of surveillance technology, and widening global inequalities.

To create a coherent global approach, several principles are worth championing:

- *Transparency and explainability:* The right to understand the logic behind AI-driven decisions that impact individual lives is vital. It fosters trust and prevents hidden biases within black box systems.
- *Responsibility and accountability:* Clear lines governing the duties of those who develop, operate, and use AI systems are indispensable. Robust frameworks enable us to address harmful outputs effectively.
- *Inclusivity and collaboration:* International forums must engage the global community of developers, ethicists, civil society, and governments from developed and developing nations. Inclusive governance is critical to ensure AI works for the benefit of all, not just a select few.
- *Agility and adaptability:* Unlike past technological disruptions, AI evolves rapidly. Global governance structures must be nimble enough to adapt, ensuring they are prepared for the unpredictable nature of innovation.

International institutions like the OECD are developing principles that nations are actively adopting. There is a trend for establishing common ethical norms. Yet, there are still challenges. It is essential to move from principles to effective regulations and to empower developing nations to contribute fully to this vital conversation.

We are at an important time in our history. Our world can benefit tremendously from thoughtfully developed AI. If it is mishandled, the negative repercussions could be far-reaching. Getting global AI governance right is imperative. Collaboration, visionary leadership, and an ethical framework are crucial as we navigate this complex yet ultimately promising challenge.

International Cooperation and Challenges

Let's consider a possible analogy for AI. A tower is built, with each brick representing a nation's vision for artificial intelligence. There is much activity in our hypothetical AI building, with some having ambitious AI plans and others cautiously laying their first stones. In our analogy, though, there is no master architect and no common language. Like the biblical story of the Tower of Babel, where people all spoke a different language, a lack of understanding undermines the collaborative effort.

In this type of scenario, international organizations like the United Nations can be useful. They can create consensus from these fragmented AI approaches and provide a forum for the global community to address its complexities. The challenges they face are as immense as the potential opportunities.

Its is probably the best time now to think of a new International AI Organization for the UN. Like, United Nations Organization for Artificial Intelligence and Innovation (UNAI) - this name reflects the focus on both AI and the innovative potential it brings, while also emphasizing the role of the UN in fostering international consensus and cooperation.

It is challenging to align divergent perspectives. Some nations prioritize AI for increasing economic growth, but their concerns primarily focus on fostering the right conditions for innovation. Others might emphasize the need for robust regulations to protect human rights and prevent potential harms like algorithmic discrimination. Navigating these differing starting points while accommodating various cultural and political ideologies requires diplomacy and patience.

Even if nations broadly agree on AI principles, enforcement presents another challenging issue. Unlike treaties for nuclear non-proliferation, for example, international AI standards might prove difficult to impose. While economic sanctions could serve as one possible punishment for rule breakers, there is a concern about effectiveness. Smaller nations with burgeoning AI capabilities may operate beyond the reach of powerful institutions.

Furthermore, a "one-size-fits-all" approach will likely fail. The context in which AI is deployed matters. Consider, for instance, the differing regulatory demands of using facial recognition to safeguard security versus social control. Balancing potential benefits with societal consequences is vastly different across cultures and political systems.

Does this mean the goal of unified global AI governance is impossible? While idealistic, the endeavor remains important and urgent. There will be many missed opportunities that occur because of poorly implemented AI standards, such as failed scientific breakthroughs and restricted knowledge-sharing that only increases divisions between nations.

International cooperation can offer a partial solution

- *Centers of excellence:* Supported by the UN, regional AI research hubs could be established to encourage innovation and collaboration, focusing on addressing shared global challenges like health care or climate change.
- *Cross-sector dialogues:* Platforms that bring together governments, industry leaders, researchers, and civil society groups become essential. These groups will not agree on everything but be able to understand diverse needs and potential conflicts early on.
- *Building bridges:* Developing nations should feel empowered to participate. Technical assistance programs and investment in education could help close the AI knowledge gap, enabling them to be full participants in shaping AI's ethical rules.

AI is not limited by national borders. Now is the time for thoughtful, inclusive collaboration at an international level. It will not be easy, but it is our best chance to create a future where AI technology fulfills its promise to create a better world for all.

AI Regulation: An Art

A traditional regulatory agency is a well-established institution with experienced inspectors. Let us consider a possible future scenario for this agency. For years, its inspectors safeguarded consumers with tested guidelines and predictable procedures. Imagine these inspectors wake up to find the world has been changed by ever-evolving algorithms and systems they might only partially understand. These inspectors might feel uneasy and uncertain about how to apply regulations to allow the AI to perform its work while still safeguarding the public.

Agencies like the Federal Trade Commission (FTC) and the National Institute of Standards and Technology (NIST) are pivotal in guiding the development and use of AI in the United States. They must promote innovation and competitiveness while shielding citizens from potential threats, such as discriminatory algorithms or privacy breaches. Yet, they cannot afford to operate with the slowness that sometimes characterizes bureaucracy.

These institutions find themselves in a difficult regulatory situation:

- *The knowledge gap:* Understanding AI technology is vital. They must learn about deep learning, neural networks, and the complexities of how biased datasets can corrupt otherwise promising applications. This ongoing learning process needs talent investment and robust partnerships with AI researchers.
- *Regulation at the pace of innovation:* AI demands faster response mechanisms than those typically used by agencies, such as those for food production. Rigid structures meant for established industries could decrease advancements instead of safeguarding them. Regulators need to be flexible, utilizing experimentation and mechanisms for regular feedback from innovators.

- *"Playing it safe" vs. encouraging bold risk:* Innovation often prospers when calculated risks are allowed. Excessive regulation risks stifling creativity and driving promising AI breakthroughs overseas. The FTC and NIST, therefore, must cultivate nuanced approaches that prioritize ethical use, transparency, and ongoing monitoring without overly dampening the research spirit. These agencies are not just adapting, they are actively creating new rules. The FTC has begun taking action against deceptive uses of AI, while NIST's Artificial Intelligence Risk Management Framework increasingly influences global thinking. It is a process of continual improvement.
- *Collaboration:* The FTC, NIST, and other relevant agencies must consistently communicate and align on guidelines, avoiding confusion for the industries they aim to regulate.
- *The "explainability" imperative:* Black box AI systems should not be allowed to operate freely. Regulators encouraging a move toward explainable AI will foster trust and help pinpoint issues early on. This requires that regulators work with research institutions and innovators.
- *International dialogue:* AI challenges do not stop at US borders. Active participation in global conversations, such as sharing best practices and coordinated action with regulatory counterparts worldwide, is vital to address common problems.

The work of US regulatory agencies in AI will influence how this transformative technology will ultimately benefit society. It is an enormous undertaking, requiring technical expertise, a bold approach, continuous learning, and a sense of urgency.

Emerging Trends in AI Governance

We live in a world where AI rules are hastily drafted. It is no longer a question of whether laws will govern AI but how they will adapt fast enough and who will set the most influential standards. We are witnesses to the interplay of state-level ambitions, international efforts, and industry-driven initiatives.

In the US, states are adopting different AI governance strategies. Some, like California and Illinois, have stricter AI regulations focused on privacy and bias mitigation. Others are primarily concerned with creating "innovation-friendly" regulatory zones, hoping to nurture a thriving AI ecosystem. This decentralized approach results in both healthy competition among states and the risk of confusion for technology companies operating across jurisdictions. It also means some important AI-related issues are getting little attention.

Internationally, bodies like the European Union are important. The proposed EU AI Act aims to be a global regulatory force, with hefty fines for those falling foul of its rules. It places AI systems into "risk" categories, attempting to balance strict measures for potentially harmful areas (like deep fake technology) with more flexible oversight for lower-impact uses. Yet, debates continue: some critics feel the act could dampen innovation, while others believe it does not go far enough in safeguarding human rights.

Trends

Here are some important trends in AI governance:

- *Ethical frameworks become more important:* Ethics will no longer be an afterthought for AI. Calls for AI transparency and accountability are occurring at both the legislative and consumer levels. The ability to explain AI decisions is becoming nonnegotiable for wider adoption, pushing companies to invest heavily in this aspect.
- *"Soft law" is just the beginning:* Principles and codes of conduct drafted by technology giants and global institutions are well-meaning but useless without stronger regulatory backing. These early blueprints now show the way for formal governance structures with enforceable measures.
- *The public sector becomes more powerful:* The use of AI within government services is accelerating, such as in predictive policing or decision-support systems for social benefits. However, oversight gaps within these public applications pose risks. Expect closer scrutiny and regulation, particularly in areas where AI could inadvertently exacerbate inequality.

The challenge is that AI governance is difficult to manage in regards to the speed of AI innovation. Regulators must balance protecting the public while not stifling AI system development progress. It demands proactive approaches:

- *"Sandboxes" for experimentation: Sandboxes* are safe environments for experimenting with AI with strict oversight. They offer a way for industries and regulators to learn together, potentially speeding up the adoption of beneficial new technologies.
- *AI audit firms:* These organizations, staffed by specialists trained to understand algorithms and potential bias, would act as vital intermediaries between industry and regulators.
- *Technologists work in public policy agencies:* Governments must build the in-house AI expertise to engage on policies effectively. Collaboration here is important, not antagonism.

Will the disparate efforts solidify into an international AI rulebook, or will competing regional strategies ultimately destroy that idea? One thing is sure: the management of AI governance within the next decade will determine whether this technology helps create a more equitable world or enhances the societal divisions of our times.

6

PUBLIC PERCEPTION, ACCEPTANCE, AND LITERACY OF AI

INTRODUCTION

Artificial intelligence is becoming an important part of our lives, and it is promising solutions and raising complex questions. This chapter examines the vital role of good governance in shaping an equitable and responsible use of AI.

We begin by addressing public perception, why it matters and how we can shape it. We must balance critical thinking with informed optimism, fostering engagement for the best outcomes. AI literacy is essential for understanding the complex algorithms utilized in AI systems.

We then discuss multi-stakeholder collaboration. Progress only happens when diverse voices, such as engineers, activists, policymakers, and regular citizens, evaluate AI's ethical and societal ramifications. We explore the challenges of the different viewpoints and the realities of implementing change.

A well-functioning, inclusive AI ecosystem requires work for successful implementation. It can be a contentious process. Yet, we need these efforts to build trust, prevent harmful applications, and create AI that truly serves society. Our future depends on how we collectively handle this powerful technology.

POLICY PERCEPTION, ACCEPTANCE, AND LITERACY

In the early days of the Internet, before everyone had an email address, there were numerous articles in newspapers and family conversations about it. We are at a similar juncture with artificial intelligence. AI governance debates occur in specialized forums, such as ethics conferences, academic publications, and government reports. However, AI success and safety will not be achieved at this high level alone. We must not only consider expert opinion but bring the conversation about AI to the public.

Why does perception matter? Mistrust of AI could prevent some of its most beneficial applications. AI-powered medical diagnostics are a good example: even a slightly better accuracy rate may save many lives, yet public fears about privacy or algorithm errors could stifle widespread adoption. Conversely, an uncritical view of AI could create vulnerabilities: people may become too reliant on recommendations by systems driven by biased datasets or with little true understanding of a situation. For AI to serve society well, a critical "middle ground" of engagement is required.

How do we go about changing public perception? We require a multifaceted approach:

- *AI literacy is essential*: Just as we do not expect everyone to be a computer programmer, we do not need everyone to be an AI engineer. Yet, a rudimentary understanding of how algorithms work, how they learn from data, and how they can fail should be considered foundational knowledge in the digital age. This education begins in schools and extends to easily accessible resources for adults.
- *Explainability over "magic:"* Companies touting their AI inventions must avoid making AI systems mysterious. Demystifying AI by explaining how it functions builds trust. Initiatives to reveal data sources and how systems generate output are essential. Explaining the system will not require everyone to become an AI expert, but it will empower people to ask better questions.
- *Show, do not just tell:* Case studies highlighting AI's successes and failures effectively communicate its benefits and possible problems. Stories of AI tools addressing health care disparities or speeding up climate research are useful for explanations to the general public.
- *Governments should be involved:* Public sector bodies using AI, whether for traffic management or tax audits, have a special responsibility for transparency. Explainability requirements must be stricter for these implementations, fostering accountability and building a blueprint for how the industry should manage similar deployments.

This approach to modifying public perception of AI is not easy. There will be those who dismiss concerns as "fearmongering" and others who resist any application of AI, no matter how beneficial. Building informed public dialogue around AI represents a vital investment. An AI-literate public can support good design, sensible application, and robust oversight of AI.

Conversations about AI limitations and societal implications should become commonplace. As we discuss complex technologies, people will engage in critical thinking and this will create a stronger understanding of how to best use AI. If we can have these types of conversations, AI governance has a far better chance of succeeding, ultimately empowering humans, not diminishing us.

Public Perception and Acceptance of AI

AI is being used in our everyday lives. From the voice assistants on our phones to the algorithms curating our online experiences, AI quietly shapes our world. These daily interactions were initially developed using a mix of awe, hope, and skepticism. To better understand why some people welcome AI while others do not, we must consider the differing points of view. Here are some views of AI:

- *AI Optimists:* For these people, AI is a tool with unmatched power to improve lives globally. They advocate for proactive policy, ethical design, and robust measures to address risks like job displacement. The AI Optimists see self-driving cars as a way to prevent traffic fatalities and believe AI-driven drug discovery can find cures for diseases.

- *The Hesitant Majority:* A large portion of the public falls into this category. While excited by the possibilities of AI, their enthusiasm is tempered with concerns. Will AI-powered hiring tools perpetuate bias? Can these complex systems ever be flawlessly secure? Will deepfakes erode trust in what we see and hear? They are not completely against AI, but they want transparency before giving their full support.

- *The Deeply Skeptical:* From fears of a machine uprising like that found in science fiction movies to concerns about manipulative surveillance regimes, this group harbors the deepest doubts. They question the motivations behind AI development, seeing it as primarily driven by corporate profit or vested interests. For them, AI's dangers outweigh its potential benefits.

- *People who view AI as savior or destroyer:* These people often think of AI as an existential-level threat (much like the fictional self-aware "Terminators" from the movies). These narratives fuel deep-seated fear and an "us vs. them" view of technology's advancement. They can also believe in the idea of "AI-as-a-miracle." For example, they can believe there is a way to create a flawless digital assistant solving global problems that humans alone could not address. These extreme ideas create unrealistic expectations, concealing the hard work, failures, and ethical issues connected with AI implementations.

- *People who are concerned with AI robots:* Often, AI in movies has a physical presence, such as a robotic body, disembodied voice with a personality, or a clever hologram. While this is amusing in movies, it contributes to an overly literal association of AI with a humanoid form. Most of the influential AI, from recommendation algorithms to stock-trading systems, does not come in these forms.

- *Media influencers:* Media rarely focuses on the gradual, sometimes subtle ways AI influences our lives. This lack of representation of commonplace AI, such as the search engine you depend on or the spam filter sorting your inbox, makes it less visible and more challenging to discuss and regulate. Because the media does not focus on this type of AI, the public may believe that important AI applications are in the distant future, ignoring that it is already subtly influencing their daily choices.

Understanding these diverging perspectives is critical because public opinion is powerful. A fearful public could slow progress, with calls for AI bans that may stop important AI-driven efforts to combat climate change or manage health care problems.

Uncritical acceptance of all AI systems is equally concerning. It can result in reckless application, unseen biases, and a gradual erosion of privacy. The goal, therefore, is to build informed discourse, where realistic assessments of both the positive and negative impacts shape how AI evolves.

We can shape public perception responsibly. Consider the following:

- *Fact over fiction:* Sensational news headlines and Hollywood's depiction of AI do little to educate. Fact-based initiatives highlighting AI's role in addressing societal issues (from improving transportation safety to finding solutions for rare diseases) can show the actual power and utility of AI.
- *The creators' responsibility:* Scientists and AI developers should engage in proactive outreach aimed at filmmakers and writers. Collaboration and consultation could inject greater realism into fictional AI, preventing harmful misconceptions from becoming part of the public's perception.
- *Media literacy revisited:* Education efforts around AI literacy should not neglect pop culture analysis. Students need the tools to dissect common tropes in AI storytelling, encouraging them to see media depictions as entertainment, not accurate predictors.
- *Addressing the skills gap:* The fear of AI-triggered mass unemployment is not without foundation. Programs targeted at retraining and upskilling workers for jobs alongside AI must become widespread, easing the anxiety associated with technological shifts. This approach transforms a source of worry into a promise of new opportunities.
- *Fairness:* Stories of biased facial recognition systems or automated loan rejections that disproportionately harm minorities reveal the potential for injustice within the technology. These failures must be discussed so that algorithmic can incorporate fairness and remind developers that AI will not automatically create a better world because it is "advanced."
- *Special AI resources:* Public discourse could be aided by popular AI resources. These could explain how real-world AI systems are crafted, discuss the data used, and reveal the existing limitations and uncertainties.

Shaping public perception of AI is not about creating relentless propaganda or silencing criticism. Instead, it requires transparency, honesty, and collaboration between technologists, policy experts, and the public. This shared conversation must start early. It is as fundamental as educating a new generation in AI literacy.

AI holds enormous potential, yet public concerns cannot be dismissed as simple misunderstanding. There is wisdom in a cautious approach tempered by curiosity. Making the public more aware of AI means we must address challenging questions about equity, privacy, and accountability. AI evolves quickly and can be intimidating. Pop culture provides a readily accessible way to enhance the public's understanding. We should be wary of encouraging fearfulness or an overestimation of AI abilities. By actively encouraging knowledge of how AI works, we can foster healthier engagement with AI.

Promoting AI Literacy

A basic understanding of how AI is a transformative technology is necessary for people to actively participate in using and managing AI systems. Just as mastering Internet tools empowers people around the globe, having AI literacy will allow the public to better engage with AI systems in meaningful ways. What is meant by *AI literacy*? Here are some parts of this concept:

- *Important terminology:* Learning about machine learning, neural networks, and the complexities of data-driven systems gives people a strong educational foundation. Not everyone needs a computer science degree to understand AI systems. Age-appropriate explanations focused on how AI operates, alongside its limitations, is an excellent first step.
- *Realistic ideas about AI:* Most individuals encounter AI in everyday life by using company AI applications, like chatbots or special algorithms. Literacy efforts must highlight how AI is subtly embedded in daily services, from curated social media feeds to targeted advertisements to AI-assisted driving features. This part of AI literacy encourages an awareness about the real uses of AI.
- *Demystifying AI:* The idea that AI "magically" gains intelligence is not just wrong; it is harmful. An essential part of literacy is helping people understand that AI learns from data and encouraging them to ask questions. Where does that data come from? Is it representative? Does it have embedded biases? These critical thinking skills are crucial for an enlightened relationship with AI.
- *AI is not infallible:* Literacy programs need to be honest about AI's potential and its shortcomings. Understanding failures, like misidentifications in facial recognition or AI tools inadvertently promoting harmful content, is vital for avoiding both naive optimism and total distrust. We need informed advocates to shape the development of AI systems.

How do we achieve AI literacy? We must incorporate an understanding of AI in the educational system. Here are some considerations:

- *Start early, teach often:* Early education programs should include basic AI concepts in age-appropriate ways. This should include coding classes and playful and accessible demonstrations across curricula. AI education should continue into adulthood via community libraries, online resources, and accessible programs targeted at adults.

■ *Governments should help:* Just as public service campaigns promote basic safety or cybersecurity awareness, a commitment to educating about AI on a wide scale signals its importance. Simple explanations for how AI tools help diagnose diseases can be placed at health care offices. There could also be initiatives focused on helping elderly communities utilize AI-powered services.

■ *Collaboration is important:* This transformation in literacy relies on partnerships. Technology companies must invest in sustained educational resources. Schools need to collaborate with those designing AI systems to bring real-world examples into the classroom. Civil society groups will play a key role in reaching underserved communities and pushing back against inaccessible, overly technical presentations.

AI literacy should not create fear or unreasonable optimism. It should assist citizens so they are capable of asking informed questions, participating in policy debates, and advocating for AI that aligns with society's goals. In a world increasingly defined by algorithms, the most impactful investment we can make is in an AI-informed populace. We need to ensure that an educated public understands how to engage with AI systems in a meaningful way and suggest improvements to them.

Making AI Literacy the Norm

Serious challenges often require bold solutions. When it comes to AI literacy, we must expand both the scope and the methods of knowledge dissemination. Here are a few areas for consideration.

Digital literacy frameworks

Digital literacy programs aimed at teaching online critical thinking skills, responsible Internet use, and source evaluation are already used in schools and community groups. AI literacy initiatives should be placed alongside these established ones. Consider the following:

■ modules addressing algorithmic bias, helping students and adults alike recognize how search results or recommendations are not neutral

■ lessons on how large datasets can skew AI outcomes, teaching the public to challenge claims relying on massive numbers without context

■ accessible content teaching responsible data practices; simple actions individuals can take to curate their own data footprint and be conscious of AI systems learning from their actions

Using popular science to teach AI

Formal education systems take time to adapt, but we cannot wait decades for widespread AI literacy. Media partnerships aimed at explaining AI concepts in accessible ways could have a big impact. Some possible ways to use popular science outlets to teach AI are as follows:

- short online animations demystifying deep learning, using metaphors and real-world examples to break down complex ideas
- segments on local news reports explaining how AI works, such as revealing how AI assists city planners with traffic or exploring how a hospital uses AI-powered tools
- collaboration with museums on hands-on, interactive exhibits, letting kids and adults experiment with simple machine learning concepts in a non-intimidating way

Upskilling for the workforce

Today's jobs might look very different in a decade due to AI. AI literacy should not be solely about individual understanding, but also about workforce preparedness. Initiatives for reskilling and upskilling could play a significant role. Upskilling involves the following approaches:

- Industry-specific programs should be designed by employers and relevant unions to provide practical knowledge of how AI functions within that particular field.
- Targeted resources should be designed for those most at risk of displacement, empowering them to leverage AI as a tool rather than fear its advances.
- Companies should be incentivized to provide regular and understandable updates to employees about how AI systems are being deployed, building a level of trust and comfort with a changing environment.

Gamification and AI-powered learning

AI itself could be a powerful tool to boost AI literacy. Consider the following:

- games and simulations that let users understand biases in AI models, demonstrating how data shapes results
- personalized learning platforms tailored to the individual's starting point, making AI literacy an interactive and engaging process
- adaptive systems that recognize when a learner may be struggling with a concept and adjusts the instruction method appropriately

Challenges and a caveat

Pervasive AI literacy will not occur without problems. It is imperative that marginalized communities do not miss out on this essential understanding. Moreover, AI is an ever-evolving field; literacy efforts need regular updates and must be agile enough to handle continuous breakthroughs and the ethical challenges they create.

It is important to note that an initiative for broad, inclusive AI literacy is not meant to imply the public needs to become AI experts. The goal is to demystify AI and empower regular people to make good decisions. Building AI literacy is not a threat to those creating this technology, but rather creates a more capable and engaged audience for their advancements. A society truly ready for a future full of AI systems needs every stakeholder to be consciously and responsibly involved in its development.

When imagining technological progress, some people might think of secluded labs filled with brilliant scientists working diligently on a new AI system. This is a romantic idea, but inaccurate when it comes to the development of AI. Today, building, understanding, and responsibly governing AI involves many people, not just computer scientists.

Multi-Stakeholder Initiatives for AI Development

Multi-stakeholder initiatives serve to guide the development of AI. These initiatives include a wide variety of different people who collaborate on AI systems:

- *The Creators:* There are the developers, engineers, and computer scientists building the core AI systems. They possess deep technical knowledge but may be less exposed to ethical dilemmas and the real-world consequences of their work.
- *The Policymakers:* Governments at all levels must address the creation of AI-relevant regulations, from city ordinances to international treaties. Balancing innovation with social good is their challenge.
- *The Affected:* Civil society groups, representing ordinary citizens, potentially marginalized groups, and users affected by AI implementations are powerful and can call attention to equity and human rights.
- *Industry Leaders:* Companies deploying AI have enormous influence and practical concerns, such as compliance issues, investment demands, and profitability.
- *Academics and Ethicists:* These independent experts examine AI's implications, often highlighting societal challenges overlooked by the excitement around technical feasibility. Philosophers, social scientists, and technologists have diverse theoretical perspectives.

Multi-stakeholder initiatives can take many forms. Global, formal entities like the Partnership on AI and grassroots citizen groups can have an important influence on AI development. Why are these types of groups necessary? Please consider the following:

- *AI reflects the ideas of its creators:* Building diverse teams from the outset leads to greater sensitivity to bias in systems. When different kinds of people can influence the AI system design, there is a better chance of creating positive real-world outcomes. For example, consider how different facial recognition technology could be if, early on, its design included people underrepresented in many of the initial development datasets.

- *AI can help avoid bias:* Collaboration prevents narrow views from shaping the AI system. Experts may not understand all of the influence their system could have on the public. Privacy and civil rights groups can question whether an AI solution that works is worth deploying at all based on values the creators have not necessarily considered.
- *Collaboration can help build informed policies:* Government entities struggle to keep up with rapid AI development. Without strong channels of communication with developers, industry practitioners, and communities, policies risk being poorly targeted or obsolete.
- *Public trust is vital to success:* Public acceptance of AI depends on good communication. People are far more likely to embrace advancements if they know their concerns and priorities have at least been considered through multi-stakeholder engagement. Lack of inclusion results in a trust deficit that can stop even well-intentioned initiatives.

Collaboration comes with its own challenges:

- *Balancing power:* Large technology companies often control the most resources. Will these companies prevent less affluent parties' views from being included?
- *Talk vs. real change:* Multi-stakeholder platforms often excel at identifying problems and outlining goals. However, there is a risk of only talking about issues and never resolving them. The ideals from collaborative discussions on AI must be put into actionable policies or changes in practices by organizations with power.
- *Speed issues:* Consensus, by its very nature, does not occur quickly, and this can have an effect on rapidly evolving AI systems. Efforts that are seen as overly bureaucratic risk failing those they aim to support, as technology outpaces outdated recommendations.

While not always idyllic, multi-stakeholder engagement around AI remains an absolute necessity. We must continue to create, iterate on, and strengthen collaborative efforts where conflicting priorities coexist to shape AI systems that prioritize progress and ethics. Collaborative efforts can be useful in helping the public engage with well-developed, ethical AI systems.

Effective Collaboration

The idea of bringing together technology companies, policymakers, concerned citizens, and ethicists to guide AI development has an undeniable appeal. Yet, too often, these collaborative efforts only result in pronouncements and statements that do not have much effect. It is time for us to consider how these collaborative efforts can truly guide the development of ethical AI systems.

We should avoid the idea that simply getting opposing parties seated together to discuss AI challenges can result in constructive outcomes. Like everything meaningful involving AI, the collaboration process will be complex.

The good news is that these discussions show that some progress is being made. These collaborations should address a few vital questions.

Does power sharing exist?

AI shapes power dynamics just as much as they shape it. A discussion involving very wealthy industry leaders, government representatives, and some community group figures does not automatically result in collaboration. True collaboration means creating mechanisms whereby those advocating for marginalized communities, for example, possess the means to help make and challenge decisions. All groups would have veto power over projects or budgets for research focused on addressing biases that have demonstrable real-world impact. All parties need to share power in the development of ethical guidelines.

How to move from discussion to action

Collaborative spaces can identify crucial ethical considerations, but then serious decisions must be made. Without clear pathways to creating change tied to the work of AI regulatory bodies, corporate policy adjustments, and grassroots efforts to empower ordinary citizens, this work is pointless. The public may even become disheartened. Can these alliances influence technology hiring practices to increase diversity among creators? Could joint projects exist between activists and companies to test problematic AI systems before deployment? We need actionable plans stemming from a collective understanding of the problems.

Building a useful feedback loop

It is unrealistic to demand that everyone involved in the process agrees on every issue from the start. An inherent feature of multi-stakeholder initiatives is the potential for a project to move forward amid legitimate dissent from certain participants. However, a key aspect of trust rests on how the outcomes of deploying AI, along with unresolved concerns, are fed back into the process. If those who sound early alarms find their fears validated on the ground, that strengthens future discussions. These collaborations should not become "echo chambers," having every participant express the same viewpoint. That type of process will fail to promote collective learning.

Participants cannot be entirely neutral

Unfortunately, not all multi-stakeholder initiatives are apolitical or motivated purely by good intentions. Some participants actively oppose stronger AI regulation, seeking to maintain an unchecked environment for rapid corporate progress. This does not invalidate attempts at collaboration, but it underscores that success can be defined differently. Instead of forcing artificial consensus, a valuable function of these alliances is to reveal conflicting agendas. Transparency about who gains from a certain development path matters as much as agreements no one is obligated to follow.

The Future of AI in Global Crisis Management and Human Enhancement

As we advance into the 21st century, the potential of artificial intelligence to address global challenges and enhance human capabilities is becoming increasingly apparent. In this section, we explore speculative yet plausible scenarios where AI can play a critical role in global crisis management and human enhancement, focusing on the necessary governance and policy frameworks to ensure these technologies are used responsibly.

AI for Predicting and Mitigating Global Conflicts

AI has the potential to significantly improve our ability to predict and mitigate the impacts of wars and conflicts. By analyzing vast datasets from satellite imagery, social media, and economic indicators, AI systems can forecast potential conflicts and provide early warnings, allowing for timely diplomatic interventions. Governance frameworks must ensure these AI systems are transparent, accountable, and used ethically, avoiding misuse and respecting national sovereignty.

Predictive Models for Humanitarian Crises

AI can predict the humanitarian aftermath of conflicts, including famine, displacement, and healthcare crises. Machine learning models can analyze historical data and current conditions to forecast food shortages, refugee flows, and disease outbreaks. Effective governance will ensure these predictions are used to mobilize resources efficiently and ethically, preventing suffering while respecting human rights.

Supply Chain Optimization for Relief Efforts

In post-conflict scenarios, AI-driven logistics platforms can optimize the distribution of aid. These systems can dynamically adjust routes and inventory based on real-time data, ensuring that aid reaches those in need efficiently. Policy frameworks should mandate transparency and accountability in these AI systems to prevent corruption and ensure equitable distribution of resources.

AI in Post-Conflict Reconstruction

AI can assist in post-conflict reconstruction by analyzing damage and prioritizing rebuilding efforts. AI can assess infrastructure damage and monitor reconstruction progress using drones and satellite imagery. Policies must ensure that these AI applications promote sustainable and resilient rebuilding efforts, with strict oversight to prevent exploitation and ensure inclusivity.

Ethical Considerations and Global Cooperation

Implementing AI for conflict prediction and management raises significant ethical considerations. Governance frameworks must ensure these AI systems' transparency, accountability, and fairness. International cooperation will be essential to develop standardized regulations and ethical guidelines, ensuring that AI serves humanitarian purposes and respects the sovereignty of nations.

The Evolution of AI-Enhanced Human Capabilities

Integrating AI with neural interfaces represents a significant advancement in enhancing human intelligence and sensory capabilities. These technologies could improve problem-solving skills, learning speeds, and memory retention. However, robust governance frameworks will be required to address ethical and privacy concerns, ensuring these enhancements are used responsibly and do not exacerbate social inequalities.

AI-Enhanced Creativity

AI transforms the creative landscape by collaborating with humans to generate novel art, music, and literature. These collaborations can lead to innovative works that blend human creativity with computational power. Governance and policy frameworks must address authorship, intellectual property, and the value of human creativity, ensuring that AI-enhanced creativity enhances rather than diminishes the artistic process.

AI and the Future of Autonomous Decision-Making

AI's potential in autonomous decision-making systems could revolutionize infrastructure management and personalized healthcare. Future AI could autonomously manage urban infrastructures and provide real-time, personalized healthcare solutions. Governance policies must ensure these systems are transparent, accountable, and ethically designed to protect individual rights and public welfare.

AI-Driven Global Governance

AI could drive global governance structures to address global challenges such as climate change, pandemics, and geopolitical conflicts. These systems could analyze vast amounts of data to predict and mitigate crises, coordinating international responses efficiently. Robust international policies and cooperation will be essential to ensure these AI-driven governance systems are fair, transparent, and inclusive, promoting global stability and equity.

With new advancements, it is less about whether they are "good" or "bad" and more about recognizing how they shift power distributions. Multistakeholder efforts will not make that struggle disappear, but they become indispensable if the aim is to prevent power from residing solely with a few participants. In this sense, a successful initiative does not necessarily yield instant gratification. Real value rests in training society to practice an ongoing, sometimes confrontational, process of collective deliberation before allowing AI systems to be implemented.

There is no roadmap for how to best develop AI systems, and so we must create one. We owe it to ourselves to treat these efforts with seriousness and scrutiny befitting something that will significantly influence the future. That means celebrating small victories as well as discussing failures openly. This, ultimately, helps foster more informed choices, not just by those designing AI systems but by citizens facing their impact on daily life.

7

AI RISK FACTORS AND HOW TO MITIGATE THEM

INTRODUCTION

This chapter examines the multifaceted risks associated with AI and the strategies for mitigating them. We will consider AI's potential pitfalls, acknowledging that the true dangers of AI often lie not in large catastrophes but in subtle ways it can gradually reshape our world, often unnoticed until it is too late.

The analogy of "technological kudzu" aptly encapsulates AI's pervasive growth and potential to unbalance our societal ecosystems. We will consider AI risks, which range from poisoned information streams to the unchecked proliferation of "smart" systems across different environments. There is a need for a holistic and proactive approach to AI development, balancing the enthusiasm for AI's benefits with the critical work of identifying its potential harms and ethical dilemmas.

We address the role of education, retraining, and robust social safety nets in preparing for an AI-transformed future, highlighting the importance of a comprehensive educational approach and a visionary social safety net system. We will address the importance of understanding and managing AI's ethical implications, emphasizing that ethical AI development is not a static goal but a dynamic, ongoing process requiring vigilance and adaptability.

We will examine some real-world case studies that illustrate successful and challenging AI risk mitigation scenarios. These stories provide valuable insights into the complexities of ethical AI implementation and the importance of continuous scrutiny and improvement.

As we forecast future trends and challenges in AI, readers should feel encouraged to engage with AI risks and their mitigation strategies. We must act with intelligent optimism, urging proactive navigation of AI's complexities to unlock its immense potential while safeguarding against the harms that might not be as overt as those in sensational stories but could be far more corrosive over time.

In this chapter, we will approach the topic with informed awareness and preparedness. We, as part of a global community, can manage AI with foresight and responsibility, ensuring that this transformative technology serves as a force for good, enhancing human life while preserving our core values and society.

AI RISK AND MITIGATION

There is much fear and confusion about AI in the media. It is easy to buy into that fear. As with any transformative technology, artificial intelligence has two paths: the one we actively design and the one that comes into being due to neglect. In many ways, the biggest AI risks are not the things that go spectacularly wrong. They are the small things that go unnoticed until they cause considerable damage.

AI can be thought of as a kind of "technological kudzu," that tenacious vine that can destroy entire landscapes. AI is not a single "organism" but a large system of interwoven technologies. Its growth is fueled by data, expanding into every corner of our lives, including health care, transport, and communications. It offers unprecedented problem-solving potential, but if left unchecked, it can stop growth and create an unbalanced system.

When AI creates news feeds without safeguards for accuracy, we all suffer from poor information. When "smart" systems, designed in one environment, are haphazardly adopted across borders, vulnerable people suffer most. Even well-intentioned AI models can become "black boxes" if not rigorously monitored. AI systems can suddenly begin automating inequity they were never meant to create. This is not a call to halt progress. We can and should be as eager for AI's benefits. This means acknowledging that true, lasting progress needs both incredible breakthroughs and the hard work of dealing with code, rethinking who designs our AIs and how those systems work for everyone.

Our regulations, shaped for different kinds of technology, cannot address the needs of AI. Traditional safeguards involving people will not suffice when AI makes numerous decisions faster than any human alive. International competition adds to this problematic scenario, fueled by the race to develop AI without regard for the standards of safety and ethics. We need a new playbook; this chapter will not offer all the answers, but it promises to illuminate the questions we dare not ignore any longer. By proactively addressing the risks of AI, we can safely utilize the incredibly good AI offers and safeguard ourselves from harm.

Education, Retraining, and Social Safety Nets

Artificial intelligence will change the nature of work as we know it. Some transformations are apparent, such as the assembly line worker competing with ever-efficient robots and the accountant ceding routine tasks to automated software. These are the stories that dominate the headlines and stoke fear. The deeper, less easily understood implications are potentially even more disruptive. Consider how AI may not eliminate jobs entirely but fundamentally alter their functions. The customer service representative now relies heavily on an AI-powered knowledge base, suggesting responses and classifying customer moods. The radiologist consults AI tools for anomaly detection, supplementing their diagnostic expertise. Suddenly, traditional knowledge is not obsolete, but neither is it sufficient.

This creates a profound educational challenge. Mastering specific AI tools is far less potent than understanding how those tools work. Can a customer service worker tell why an AI suggestion might be inappropriate? Can a radiologist understand when the algorithm identifies something legitimately concerning versus mistaking a normal variation for a problem? There is a need for more education. Historians can identify data bias in historical reporting. Artists can harness AI more powerfully when they understand its capabilities and limitations. The non-technical disciplines may be helpful when engaging with AI systems.

Here is where we risk making a dangerous mistake. We may hurry to retrain displaced workers as coders or data scientists without fully addressing the level of critical thinking about AI itself. This is much like the difference between teaching someone to fish and how to assess the health of the entire lake.

Alongside improved AI educational initiatives, we desperately need a social safety net. Traditional notions of worker aid must adapt to ensure those impacted by AI do not suffer needlessly. A truly visionary approach does not only consider damage control. We can envision unemployment support as a stopgap and as seed funding for promising ideas. What if the factory worker laid off, thanks to their front-line knowledge, has a brilliant notion for process optimization with existing AI tools? What if the accountant, freed from drudgery, could design financial literacy programs for communities historically underserved?

A strong social safety net fosters experimentation and risk-taking, precisely the qualities hurt by job insecurity. In an economy transformed by AI, it is a prerequisite for sustained innovation. Imagine an environment where people are not simply retrained to fit specific jobs but instead are empowered to co-create the jobs that will be needed with AI as a powerful collaborator.

The path to success will not be clear-cut or easy. They will require bold experimentation, collaboration across unlikely disciplines, and an understanding that adaptability is as vital a skill in an era of rapid change as technical knowledge. The issue is whether we will invest in these ideas, not only because they are pragmatic but also because they align with our highest goals as a society: using human ingenuity to ensure technology enriches rather than erodes our economic and social well-being.

Understanding and Managing AI's Ethical Implications

We often have a simplistic idea of ethical AI development: malevolent actors seeking to harm and well-meaning creators ensuring their technology is unbiased and fair. Reality, however, is far more complex. The most troubling ethical breaches can stem from noble goals built on faulty assumptions and challenges we did not even know we had. Consider an AI model designed to help manage prison parole decisions that identifies potentially high-risk offenders. It sounds prudent, but if trained on historical data, it perpetuates the deeply rooted racial biases of the justice system. No one wanted to build a racist algorithm, yet the inherent limitations of the data led to exactly that.

This calls for a radical shift in how we conceive of AI ethics. It is not enough to debug an algorithm once for fairness. AI systems interact with a dynamic world. The data becomes outdated, societal shifts render our original assumptions obsolete, and subtle errors or new applications create unintended consequences. This is the concept of *drift*. An initially well-designed AI system can experience ethical quandaries simply because the world is not static. Ethics must be a part of an ongoing process. Companies must develop monitoring and auditing systems to find and address these unexpected side effects of seemingly beneficial technology.

This highlights a problem deeper than just flawed code. Bias emerges from the worldviews and backgrounds of those who conceive of, design, and train AI systems. The technology industry has a well-documented diversity problem, so even well-meaning projects lack the perspectives needed to see potential issues. This problem cannot be fixed by a few ethics lectures for engineers. True diversity of thought is needed at every stage of the process, from initial problem definition to data collection to ongoing deployment. We need artists, social scientists, and people representing various communities who will be affected as collaborators in shaping the technology.

What happens when, despite our best efforts, harm occurs? Accountability cannot solely rely on the court system to resolve matters. Lawsuits are reactive, expensive, and may not provide justice. A system built on a biased dataset might escape litigation despite clearly perpetuating social inequality. We need robust industry-wide standards so companies cannot claim naivete, shift blame, or pretend the code is too complex to be fixed. More importantly, we urgently need accessible mechanisms for those affected by AI-driven decisions to seek redress. Can the person denied parole challenge the system, even if they cannot identify a specific coding error? Can a community subject to heavy surveillance based on predictive policing models demand explanations and safeguards?

AI regulation and policy is an entirely new area. It will not be perfect, and the solutions will not always be neat or convenient. Yet to abdicate responsibility is to surrender the promise of AI to those who might profit handsomely from unchecked algorithmic power. Innovation that enriches human experience occurs when we treat AI as a powerful tool that must be wielded with an equal focus on technology and ethical vigilance. This is a challenge we simply cannot afford to ignore.

AI's Entanglement with Everything

Many people have deceptive and uninformed ideas about the implementation of AI systems, such as those seen in science fiction movies like *Terminator*. Right now, AI is not an external force but one quietly working in common systems we rely on, the decisions we make, and the knowledge we trust. The risk is not superintelligence surpassing us as much as an invisible complexity undermining us.

Let's consider the idea of the *smart city*, which is an AI-dominated metropolis often considered the future of urban development. In smart cities, there are sensors, self-driving cars, and AI-optimized traffic flow. AI might also monitor energy use to identify areas of need and guide resource allocation after disasters, all without a citizen being fully aware that it is influencing those outcomes. That kind of power carries profound and often overlooked risks.

Consider a seemingly "harmless" case, such as AI sorting online job applications. This sounds like a benevolent use of AI, removing human bias and evaluating applicants solely on skills. If the job sorting AI was trained on existing data, it may have learned what "successful" looked like historically, which may simply solidify discriminatory hiring patterns. The blame for this problem is challenging to assign. Was it the AI developer, the company deploying it, or the deeply flawed hiring ecosystem that led the machine to interpret the data in that way?

Once AI is placed into larger decision-making systems, predicting everything from crime hotspots to where medical research grants go, disentangling good intentions from problematic real-world impacts becomes incredibly challenging. Who holds the ultimate accountability? When unintended consequences surface, what is the fix, tweaking an algorithm or overhauling societal structures the AI simply learned to copy? These complexities trouble AI experts.

Addressing AI risk is distinct from other technology dangers. Unlike hacking threats, there may be no prominent "bad actor." Even well-intentioned AI models can become harmful tools if the social systems they are inserted into lack fundamental fairness, be it biased data, lack of oversight, or simply misjudging the speed of societal changes AI accelerates.

Solutions to this problem should not include a global ban on AI systems. AI offers amazing potential, and we should not demonize the technology itself. We must change how we approach progress. Just as we do not lay power lines haphazardly and hope for the best, we need a new mindset focused on safety, transparency, and continual oversight of these increasingly potent systems. We must safeguard innovation. This requires a level of maturity and public understanding of AI that most societies have not attained yet. Attaining it may be the most urgent challenge, as we may find the risks outpace our awareness of them.

Principles-Based AI Regulation

No one would regulate the automobile industry with laws designed for horse-drawn carriages. You would have rules on the length of whips, fines for startling livestock, and all sorts of misplaced measures. The same sort of regulatory challenges is involved in managing AI. Trying to control one of our era's most dynamic technological forces with legislation written before it was invented is dangerous.

Lawmakers are not lazy or inept, but the pace of change and how AI challenges our definitions of things we thought we knew creates numerous problems. When deepfakes emerged, specific bans were drafted, aiming at the tools used to create false digital constructions of reality. By the time those laws were enacted, the technology for creating such content had evolved. Laws addressing one AI technique usually result in those with nefarious intent just utilizing a newer technology.

This challenging problem of requiring dynamic regulation requires a fundamental shift in how technology rules are managed. One example to consider is the implementation of rules developed to address early Internet use when everything from scams to content warnings was new. No law listed every bad thing someone could do online. Instead, principles emerged as protections against fraud, safeguards for children, and a growing understanding that information, even if freely transmitted, was not always harmless. That established a framework that was not flawless but adaptable. We need that principle-based thinking for AI regulation.

What might these AI principles look like? *Transparency* is an explanation of how a "smart" system reached its conclusion, and it is vital for credit applications or facial recognition. Fairness prevents AI from amplifying existing discrimination, not through wishful thinking but through mandatory bias audits embedded in its development and use. Human oversight should also be included in the principles. A medical AI tool suggesting unusual diagnoses cannot replace a doctor, but it needs ongoing evaluation to see if its reasoning remains logical and information based.

The "how" of this sort of work is complex. Industry bodies, independent regulatory agencies, and new types of AI auditors can peer inside algorithmic black boxes, but none of these can be created very quickly. We should not use rigid rules specific to today's technology for AI, or we will increase the risks from such systems.

Some argue this approach limits innovation. Should not the focus be on unleashing AI's full potential? It is the classic dilemma, pitting safety against unfettered progress. History is full of examples of how people managed this type of debate. Did seatbelt or food safety laws stop cars from being produced or close grocery stores? They did not, and instead, they fostered public trust and, in the long run, created an environment where those technologies could safely bring their benefits to more people. It is time to apply that long-term thinking to our newest revolution. AI is too potent to leave ungoverned for long.

Questions of AI Ownership

Let's consider the democratizing power of technology through a comparison. The lone genius myth is very common, and many people know stories about a child who developed a brilliant idea that disrupted industries. Now think about AI (the kinds of big, complex models making news) and consider its uses, from generating believable text to analyzing vast medical datasets for patterns no human can identify. These two scenarios are very different.

Ownership in this example is not as simple as who holds the software patent. It is about data control, the enormous datasets of images, text, and medical records AI consumes as material for its learning. It is about computing power, a near-monopoly held by a handful of technology giants with immense hardware no start-up or university can afford. Suddenly, those old anxieties about big corporations stifling competition start having a terrifying, futuristic dimension.

Let's consider this from another perspective. An independent researcher may have a groundbreaking theory about early diagnosis of a specific disease. Without access to massive, diverse patient data for AI model training, that idea will not progress to a tool in hospitals. If their only path is partnering with a corporation dominating this field, their innovation may be subject to that company's profit motives or strategic alliances. If such models increasingly power critical fields like health and research, how much is truly private intellectual property (IP) versus a resource with public good ramifications? This is an incredibly challenging problem.

In addition, the challenge becomes greater when we incorporate many different countries into the problem. Countries without those technology companies are at a disadvantage. Do they adopt models created elsewhere, with little insight into how they came to be, perhaps trained on data with values or demographics utterly mismatched to their population? Is the alternative building national data troves, and starting privacy debates? The strong technology companies shape more than what apps we use. They also manage the tools of global policy in health, economics, or even military AI applications, where lack of transparency becomes a national security issue.

Solutions are not easy. It is tempting to call for breakup of AI's big technology companies, thereby enforcing public data availability. This approach comes with risks. We do not want a scenario where innovation cannot occur. Making data access "free" does nothing if a significant imbalance exists in who has the processing capacity and talent to use it well.

A reasoned approach may require unprecedented public-private cooperation or an entirely new type of institution that pools resources internationally but with ethical, equitable standards in the framework. It is unlikely that simple market competition will be able to successfully address this challenge. These issues of AI ownership could well determine who sets the standards for what these intelligent systems can and should do in the coming decade.

These issues of AI ownership could well determine who sets the standards for what these intelligent systems can and should do in the decades to come. That brings us to the primary dilemmas:

- *"Data is the new oil:"* If control of AI models is predicated on who controls the vast amount of data (or "oil") on which they train, do we simply repeat the power imbalances of the past? Those owning the data have leverage, not just commercially, but to decide what is worth teaching their AIs. In turn, that data's biases, gaps, and priorities will affect everything created from it.
- *Limited innovation:* We must be cautious to not make innovation too great of a challenge. There are two types of innovation: the disruptive kind, where smaller technology companies can use AI boldly in unforeseen domains, and the continuous kind, where widespread scrutiny and adaptation of models makes them better on issues of fairness and societal good.
- *Geopolitical limits:* Imagine a world where medical AI breakthroughs rely on access to models largely fueled by US or Chinese patient data. This is not just about healthcare access but the values embedded in how problems are diagnosed and what conditions are prioritized. Then factor in the temptation to hoard that power. Suddenly, we may have an issue with how diseases are addressed and how scientific knowledge grows (or does not) because its tools are under political control rather than being a broadly shared, adaptable resource.

These are incredible challenges. Reasonable solutions will require serious consideration and are not likely to develop on their own. Historically, this is one of the lessons of people with power and resources. However, the cost of failing to address these issues now will only escalate, making them far more painful to manage as AI further entrenches itself in our world.

The AI Hype Cycle vs. The Work of Mitigation

AI seemed to be very popular when it was first developed, and it was perceived as a challenge to old, slow industries by promising to fix problems quickly and logically. That idea can still be found in some places, although it promotes a dangerous way of thinking about AI. This dangerous way of thinking emotionally about AI is promoted by media *hype*, or an exaggerated kind of publicity. The *hype cycle* refers to how the media introduces material over time. The real danger is not in AI systems becoming independent; it is our own desire to avoid taking the steps needed to avoid disaster.

Let's consider an example. Newspaper headlines talk about an AI writing sonnets to rival Shakespeare, and the public is disturbed about what this means for artists. In the same newspaper, there is a report about a medical AI making accurate diagnoses. That kind of attention distorts how resources get spent, how people trust (or fear) AI, and what issues lawmakers react to in a way to address this technology trend.

In this imaginary scenario, who is dealing with the actual problem? The people who are managing the problem are those data scientists who are meticulously cleaning biased data so it does not poison the AI system. The problem-solvers include the engineers who are developing protocols for when that medical AI fails. Lawyers and ethicists are managing the slow, complex work of creating rules for when a self-driving car's choice is between hitting two pedestrians or harming its own passenger. This is the challenging work that separates "AI potential" from actual AI solutions.

Let's consider another comparison between AI and a renovation show on TV, where workers fix up a dilapidated house. People watching the TV show see only a small part of the renovation of the house, and much time is spent on how the house looks after it is finished. Most of the budget and time went into unglamorous aspects of the project, such as ensuring the wiring would not cause a fire and sealing the new windows against unseen drafts. If the renovators do the work wrong, that new kitchen will be ruined in a year. It is the same principle with AI: people only learn about a small part of it through media hype, such as the exciting breakthrough that makes news. They are largely unaware of the safety checks and rigorous validation. They may not even be aware that someone had a willingness to confront the fact that sometimes, smart solutions simply are not feasible with the complexity at hand.

For true and sustained AI success, the public must see all of the process that goes into creating AI systems and protecting us from unintended harm. Let's think carefully about where we focus our attention and praise regarding AI development. The stakes are high and making "intelligent technology" truly intelligent takes far more than brief newspaper headlines acknowledge.

Let's consider a specific example: the algorithms governing your social media feeds. The media often portray this as a benevolent personalization engine ("See more of what you like, discover relevant content and a digital world tailored just for you.") The media hype is positive. In reality, there are serious concerns to address for social media algorithms. Engineers try to modify those same algorithms so they can avoid sending users misinformation, politicians are concerned about the algorithms encouraging extremism, and researchers puzzle over how to mitigate the proven mental health harms this system has on some, especially teens. It is unlikely that the developers at the social media company were determined to destroy democracy or damage teens' well-being.

Even though the initial intent was to encourage social interaction, the scale of the AI's impact creates unexpected problems. Algorithms can amplify conspiracy theories or make eating disorders appear aspirational. Correcting the results from the algorithm involves significant work and an understanding of how seemingly neutral features can lead to unexpected results. Social media algorithms are designed to prioritize the most engaging content, such as promoting what your friends share, and this data can have an influence that no single person at that company fully controls.

AI risk could involve a degradation of societal trust. Algorithm fixes are debated often, not for technical complexity, but because free speech clashes with algorithmic harm, and nobody, including the creators, has the perfect answer for how to correct the problem satisfactorily. This is an important aspect of AI implementation that is overlooked amid the hype. Attempting to create solutions may curb some benefits of AI, and people may demand accountability with no obvious individual or organization who can accept it. At the same time, AI keeps learning and changing as user behavior and content evolves.

How we address these difficult interactions between AI and society will matter far more in the long run than which lab manages to make the best chatbot. We need to carefully consider our fears, hopes, and investments regarding AI.

International Competition and Unequal Impact

We often discuss artificial intelligence in two ways: utopian dreams of global problems effortlessly solved or alarmist narratives about a winner-takes-all technology battle between superpowers. Both views fail to address a more pressing issue: The profound unease about AI arising globally, particularly in nations outside the wealthy few dominating AI space.

Let's consider a possible scenario. A developing nation facing public health struggles turns to an AI-based diagnostic tool with impressive results. The nation lacks the internal knowledge to evaluate the AI tool but assumes that the tool is useful because it works for others. Then, problems begin to surface from bias in the tool due to unrepresentative data, malfunctions causing missed diagnoses, and no recourse for those harmed since its developers are from another country. In this situation, there was no intended malice, but poor design and a lack of careful evaluation for a certain purpose.

If we scale this fictional scenario, we can imagine systems used for credit approval, crop management, and policing that are imported from other countries. This importation of AI systems creates a dependency cycle with the possibility of exploitation. Data gets exported, fueling AI built elsewhere, and then it is sold back as a product the nation never had the infrastructure to create on its own. This creates an imbalance, which breeds distrust, not of the AI itself, but of the idea of technological progress.

In our scenario, it is at this point when the situation can become dangerous. It creates a fertile ground for those hyping AI as a quick fix. In some cases, authoritarian may use AI-powered surveillance tools to ensure "safety" and governments may make unfair deals with companies prioritizing profit over ethics. These are serious problems that cause harm, and they can contribute to global instability.

Our imaginary scenario does not naturally mean we should engage in a kind of protectionism for technology where we eliminate certain nations' access. Instead, it should make us carefully consider responsible AI standards. Not engaging with ethical issues has resulted in everything from environmental pollution to child labor, and engaging in poor ethical behavior does not create global prosperity. If "bad AI" can operate without guidelines, everyone suffers: innovation no longer progresses, public trust dwindles, and conflict increases as disparities blamed on tech, not on the structures perpetuating them, intensify.

Who determines the baseline for safe, ethical AI? Should it be the AI companies dominating the technology right now? This brings us back to those worries about technology empires. Should individual nations be subject to the current, poorly developed rules for data privacy? Ethical AI guidelines may require forms of cooperation we have never managed before to ensure "bad" models and the structures enabling their unchecked export face strong deterrents. Nations will need to cooperate to ensure AI's potential is not poisoned by inequity from the start.

Our traditional policy structures are challenged by the policy needs of AI. Trade deals are usually negotiated about finished products, not a technology's underpinnings. Current policies may not be able to address the issues of how an exported AI may be implemented into local assets, such as medical devices and social media platforms. General policies will not be able to address the nature or AI or its outcomes without careful consideration.

International AI review bodies, however controversial, cannot be dismissed. These international review organizations should have the power to analyze, with some independence, what risks reside within models used for critical functions. These groups will consider questions we do not yet have answers to. Will nations allow scrutiny of their training data, which is currently kept secret for many commercial AIs? Can such a governing body remain neutral, without being unduly influenced by technology or political biases? What power to enforce its will exists? Does it have the ability to sanction nations or the power to deny the use of particularly problematic AI?

The creation of an international review board for AI acknowledges the need for a safety net in the global AI market. Even well-intentioned nations attempting to implement smart systems lack the resources for careful internal vetting. A respected international body can create a baseline to protect against predatory use of AI products or poorly designed AI systems.

Operating a successful international review organization will not be easy. There are tensions between individual national interests, corporate secrecy, and varying philosophies. Ignoring those issues will not make them disappear. However, an international board could ensure that AI is not used inappropriately, and that unequal AI access does not result in inequality. It could help to ensure that every technological advance does not become a new source of suspicion and division but rather a step toward shared prosperity.

Addressing Soft Risks from AI

While there is a natural concern about the "hard" risks of AI, such as from an AI with the ability to cause harm, we should also consider "soft" risks. Soft risks include problems like misinformation, and these risks can erode trust and societal bonds.

Misinformation is a serious challenge to public health and safety. When paired with AI-powered content generation and precision targeting on social platforms, misinformation can spread with a speed and impact previously impossible. AI itself may not be authoring those falsehoods, but it is the delivery mechanism, optimizing spread based on what triggers engagement, often outrage or fear. The AI system was not intended to be used to spread misinformation, but this occurs simply because the algorithm is working precisely as it was designed.

Another soft risk is that of personalization. Algorithms learn what gets your clicks and deliver targeted content to you that matches that criteria. While this approach seems beneficial, it can result in users only obtaining a very narrow range of information and viewpoints. By creating "echo chambers" of only the same view, the AI can make some people become more extreme in their opinions. Who is accountable when personalization becomes a problem? In this situation, as well, the AI was performing according to its intended function, and no malice was purposely included in the design of the system.

Managing soft risk can be especially challenging, as it involves policy affecting the limits of acceptable free speech. Literacy can be part of the solution, as it involves teaching the young to identify deepfakes and instilling an understanding of how AI systems function. Soft risk requires societal debate on what truths holding our communities together are worth actively safeguarding.

Humans and AI Safeguards

We discussed the idea of a single decisive human overseer of AI as our ethical safeguard. While the idea is comforting and provides someone with a clear responsibility for safety, this model will not suffice.

Let us consider an example of AI in diagnostics. Naturally, the doctor should be the final interpreter of the patient's symptoms and test results. For human doctors, though, reviewing every bit of reasoning an AI uses when it makes suggestions is unlikely. A human overseer in this example is not sufficient to ensure that the AI diagnosis is entirely correct. A collaborative decision-making system is required, and it should incorporate explainability, with the AI showing how it obtained its results. Human judgment is still vital to the diagnosis, but AI enhances this ability.

Involving a human overseer in a dynamic situation where AI is used can be a more challenging topic to consider. For example, an AI managing traffic flow makes rapid decisions, and some of these decisions will be so fast that a human cannot intervene. In these situations, we must consider the nature of the system and how intervention can enhance outcomes. Legal frameworks should take into account these types of scenarios, where responsibility relies on continuous monitoring and course correction when issues surface. AI will require entirely new accountability structures for decision-making and outcomes.

Data Bias and Mitigation Strategies

AI systems are often considered bias-free and more logical than a human. AI, however, learns from the data we feed it, and our data is filled with historical and ongoing biases. An AI trained unknowingly on biased training data does not an escape from human errors but becomes an amplifier of them.

Awareness of possible bias in the data is the first step toward progress. There are several possible ways to approach the management of bias in data:

- *De-biasing datasets:* It is impossible to erase all bias, but actively identifying and addressing how past injustices skew data is the first step in creating better information for the AI. For a medical diagnostic AI, it might mean ensuring images representing varied skin tones are used so diseases are not misdiagnosed due to skewed training data.
- *Diverse AI teams:* Teams entirely shaped by one demographic may result in bias. Technology teams should incorporate ideas of social equity so the AI will be able to provide benefits to all users.
- *Proactive transparency:* Often, the problem is not in the initial biased dataset but in how an AI evolves as it is fed new inputs. Mitigation lies in continuous audit processes, seeking out real-world situations where the AI acts unfairly not because of malice but simply because it has not encountered data reflecting novel lived experiences.

Real-World Case Studies in Successful Mitigation

Here, we will examine a few examples showing how addressing these challenges fosters AI solutions that can be trusted.

Participatory design in predictive policing

Instead of developers designing an AI to identify crime hotspots in isolation, a system was built with ongoing feedback from affected communities. This led to system revisions, such as the AI emphasizing prevention strategies (such as lighting improvements and youth outreach) over arrests. Trust in the system emerged through the active participation in creating the data for the AI.

Inclusive health care AI

A project training AI in breast cancer detection sought out historically under-represented patient data. This was not done just out of fairness but practicality: it significantly improved model accuracy. A more varied pool of samples helped it discern true disease indicators from patterns limited to one group.

"Red teaming" AI products

Inspired by cybersecurity audits, some firms employ an ethical hackers to test their systems. Independent parties actively try to find biases or loopholes in the AI design before deployment. This makes finding flaws a point of pride, not something to be feared or hidden, creating a healthier overall system for AI implementation.

It is crucial to understand that none of these examples achieved AI perfection. Instead, they model possible approaches to mitigation, including continuous scrutiny, improvements based on real-world feedback, and the willingness to hold those deploying AI accountable.

CHAPTER

8

GENERATIVE AI: CONVERSATIONAL AGENTS AND BEYOND

INTRODUCTION

In this chapter, we consider the innovative world of generative artificial intelligence. This exploration involves viewing the technology in new creative ways. Here, we will examine AI's complex algorithms as a transformative force reshaping our approach to creativity and problem-solving.

Generative AI may be thought of as a new kind of partner in the creative process. An artist with a half-formed idea or a programmer with a partially built concept can benefit from a generative AI assistant who can leverage its training on massive amounts of data to help them complete their tasks. This is a collaboration where human imagination and AI's pattern recognition and extrapolation abilities can work together.

We will consider the history of generative AI, tracing its evolution from the early days of AI through the developmental milestones that have shaped its current capabilities. This historical context provides a foundation for understanding how we have reached this point in AI's development.

We will also examine the technical underpinnings of generative AI. We consider how artificial neural networks and natural language processing form the backbone of these systems, enabling them to process and generate new content with a level of sophistication that parallels human creativity.

With a focus on applications, particularly in conversational agents like ChatGPT, we see how generative AI is revolutionizing how we interact with machines. From customer service to content creation, these systems are making human-AI interactions more intuitive, responsive, and contextually aware.

Finally, we discuss these advancements' broader societal implications and policy challenges. We must carefully consider the ethical dimensions and regulatory frameworks needed to guide the responsible development and use of these powerful technologies.

This chapter is not just a technical overview but a thoughtful exploration of how generative AI is affecting human creativity and innovation. We should engage with generative AI critically and creatively to shape a future where AI amplifies human potential rather than simply automating tasks.

GENERATIVE AI

Let's consider the act of creation, whether it is painting, writing, or inventing, as a form of problem-solving. The artist begins with an idea and then makes changes to that idea as she proceeds. That process is iterative. Mistakes are corrected, older ideas are changed and revised, and prototypes fail before the final product appears.

With generative AI, that problem-solving process acquires an assistant. When working with an AI, you can begin with a sketch, a few words, a rough image, and a bit of code with flaws. Generative AI systems, trained on large datasets, are adept at pattern recognition and extrapolation. They can take a partially developed concept and extrapolate it in surprising, intuitive ways.

This ability to interpret ideas opens up possibilities that affect how we perceive creativity. Is it collaboration with an artificial mind, or are we slowly learning to become better conductors of preexisting knowledge? Consider image generation: typing "majestic lion silhouetted against a red African sunset" into a generative AI engine yield results that rival a skilled photographer's work. The AI's ability to interpret the text in a visually pleasing way does not just democratize the act of visual creation; it challenges the idea of what an "original" image means.

Naturally, an ability generates ethical concerns. Can generative AI render writers, artists, and even coders obsolete? Will these systems flood the world with indistinguishable fakes, diluting trust in information? The answers, just like the technology itself, are still evolving. Yet, history reminds us that while initially fraught with concern, technological revolution tends to open up previously unimagined avenues once we harness it intelligently.

Generative AI could reveal the potential of students who cannot effectively convey their ideas in traditional essays. It might motivate innovations in medical diagnostics, where an AI can suggest novel paths of inquiry beyond what a human doctor might consider based on past evidence. True success will not come from viewing AI as a replacement for human ingenuity but as a tool amplifying the scope of what we can collectively achieve.

Like any tool, however, the use of generative AI requires critical thinking and responsibility. If, as Bill Gates observed, technology is often just an amplifier of underlying human tendencies, we must work diligently to ensure its applications are used for knowledge-sharing, problem-solving, and enhancing human expression, not eroding it.

Historical Evolution of Generative AI

It is easy to view generative AI as a technology that only took a short time to develop. The history of generative AI is far longer than most people realize. The core concept took shape decades ago, and it was developed by researchers who were considering questions about how machines could truly "learn."

In the early days of artificial intelligence, around the 1950s and 60s, the emphasis was on rule-based systems. These first steps were much like teaching a computer how to play chess. Each part of the program was coded with every possible move and every tactical guideline. While this approach was effective, it is inherently rigid. Could a machine "learn" in a less prescriptive way?

That question spurred the early work on neural networks, which were inspired by how our brains function. Researchers theorized that by creating layers of interconnected "neurons," much like those in the human brain, a machine might start to recognize patterns independently rather than needing a step-by-step roadmap. Early work was hampered by limited computational power. These neural networks needed considerably more data.

A milestone was reached in the 1980s with the development of Markov chains. These simple mathematical models predict the likelihood of certain events based on what came before. A Markov chain is much like a basic auto-complete: If you type "Hello, how are yo," a Markov chain might suggest the word "you" as the most probable next word in the sequence. While rudimentary, using previous elements to predict future outcomes lays the groundwork for more sophisticated generative models.

Throughout the 1990s and early 2000s, progress occurred sporadically. Machine learning as a field made progress, but early generative systems tended to be highly specialized. They might generate crude, pixelated images or basic chunks of text based on specific parameters. Impractical to implement on a broad scale, they mostly remained an academic pursuit.

Then, three events occurred. First, the exponential growth of computing power made managing enormous datasets feasible. Second, the Internet provided the significant amounts of data these systems needed in the form of images, text, and musical patterns. Finally, algorithm refinements, like the development of Generative Adversarial Networks (GANs), unlocked better ways for those artificial "neurons" to learn more efficiently.

This convergence of three important events unleashed a breakthrough in AI. Generative AI models could create human faces that were nearly impossible to distinguish from real photographs or supply paragraphs of text, rivaling student essays' sophistication. This event showed that machines were not just mimicking or automating: they were starting to extrapolate and surprise in distinctly creative ways. Importantly, these tools were democratizing and no longer confined to specialized labs.

Exploring the Underlying Technologies of Generative AI

Let's consider an example to help explain generative AI. Imagine learning a new language without any textbooks or a formal school curriculum. You might simply immerse yourself in the language by listening to foreign songs, trying to decipher street signs, and watching how people interact in markets. You start to understand patterns over time, even without knowing every grammatical rule. You gain an implicit understanding, enough to make yourself understood with a broken phrase or a gesture.

In a surprisingly similar way, the core of generative AI revolves around this notion of pattern recognition on a massive scale. At the heart of these systems lie artificial neural networks. These are not exact replicas of our own brains but rather simplified mathematical models loosely inspired by their structure. They can be thought of as layers of interconnected nodes, like a "digital spiderweb."

At first, this "digital spiderweb" does not know much, but it can "learn" from data. For example, to make a generative AI that specializes in generating realistic nature photos, we need to provide it with thousands, or millions, of actual nature photographs. With each image, the network tries to identify recurring traits. Which combinations of pixels tend to form leaves versus rocks? What color palettes appear during sunsets? This process is not about storing precise images, but rather detecting statistical regularities hidden within the data.

Another crucial part of a successful algorithm is natural language processing (NLP). When generating text beyond simple phrases, these systems need to have an understanding of grammar, context, and human sentiment. NLP techniques train AI models to analyze massive quantities of text. They learn probable word orders, the connotations of various phrases, and the way writing evokes different emotions.

It is important to note that this process involves guided trial and error. Many advanced generative AI models work like internal competitions: one part focuses on creating content, while another constantly evaluates whether the results are plausible or realistic (depending on the desired outcome). These models refine themselves through this continuous self-feedback.

This is a simplification that ignores some complexities. Specific types of neural networks are designed to handle varying tasks: some excel at identifying images and others at managing language. It is crucial to understand that these generative AI systems are not truly sentient. They cannot comprehend the nuances of human experience or develop entirely novel ideas on their own. Ultimately, they are incredibly sophisticated pattern detectors and extrapolators, able to leverage the massive amounts of data we provide them.

Chatbots and Conversational Agents

The earliest iterations of chatbots were used in customer service systems or as Web site assistants, and they consisted of pre-programmed responses.

Interactions were limited to the program's capabilities, sometimes frustrating, and did not yield the best results. If the human interacting with the chatbot did not use very specific language, the chatbot could not appropriately respond.

Conversational agents like ChatGPT and similar technologies are very different from these early chatbots. Conversational agents are in a different class because these generative AI systems have been trained to utilize nuance and context. They can identify where a dialogue ended, adjust their tone based on your specific query, and generate different creative solutions from prompts like "Write a funny poem about a clumsy programmer." The human-AI exchange becomes an evolving conversation much like that between two people.

This breakthrough has serious implications. Although AI could do simple troubleshooting or provide product information, AI-powered chatbots could also act as personalized tutors, offering step-by-step guidance to students who need help with challenging concepts. Businesses could create smarter customer service systems, understanding and addressing issues more efficiently. Users can use their voice to find relevant information rather than trying to type in the most effective search engine queries. The change in technology is a significant advancement, moving AI from simply retrieving information to synthesizing a coherent response tailored to individual needs.

Of course, progress invites a certain caution. There are legitimate concerns about these systems spreading misinformation or students becoming over-reliant on them for completing assignments. Yet, history reveals an interesting pattern: technologies that initially cause worry ultimately to reveal unforeseen possibilities once we master them. Generative AI will not replace human capabilities but is a powerful new tool to assist people.

For example, a doctor may get help from an AI researching rare conditions, locating connections in medical literature they might have missed due to the large amounts of data. Consider how language barriers may erode as these AI translators use dialects and cultural idioms seamlessly. What would fluency mean for sharing ideas globally in this conversational future?

Generative AI is a powerful tool that we should use cautiously. We should explore its implications not just as scientists or technology entrepreneurs but as educators, writers, and ordinary citizens. Just as with every powerful new tool, the onus falls on us to shape how this transformative technology evolves alongside human ingenuity, not at its expense.

Applications of Generative AI in Various Industries

Generative AI can help many more people become more creative and innovative. It can be utilized in almost every industry, whether it is revolutionizing healthcare or reshaping the way we build cities. Generative AI could prove as transformative as electricity was in its time.

Let's explore some of the most exciting possibilities of generative AI.

Medical applications

Generative AI may make it possible to advance beyond only analyzing medical images. It may make it possible to create research laboratories where scientists utilize generative AI to model potential new medications. These systems can analyze enormous molecular datasets, pinpoint promising candidate compounds, and predict possible side effects. Generative AI would not replace scientists but augment their work, speeding up breakthroughs that might save lives.

Personalized marketing

Generative AI could affect the advertising industry. Companies may soon craft tailored ad copy and entire image sets, videos, and brand voice adjustments unique to each potential customer. How can we balance such powerful targeting with ensuring privacy, fairness, and the ability to opt out?

Transforming workflows

In sectors that require data, paperwork, and pattern recognition, the impact of generative AI will be profound. Attorneys could have contracts analyzed for anomalies faster than any human team, with precedents and related statutes identified quickly. Financial analysts or insurance providers may be aided by AI in parsing risk, suggesting investments, and streamlining underwriting processes that used to take weeks. Entire processes can be redesigned based on generative AI's abilities.

Digital collaboration in creative fields

Artists, architects, and game developers have a new toolkit. An architect might experiment with novel structural variations inspired by nature and generated with a few prompts. Video game worlds could become very detailed as AI creates realistic landscapes or dynamic city environments while developers focus on game mechanics and narratives. AI can assist creators and challenge traditional definitions of authorship and originality.

Because so much transformation from generative AI is possible, we must consider some important questions:

- How do we prepare for the change generative AI introduces?
- Are existing policies on intellectual property sufficient for AI-generated artistic work?
- How do we safeguard against job displacement while training those working in affected fields to use these tools strategically?

Engaging in ethical discussions about generative AI now is crucial, not when the technology has surpassed our capacity to shape its implications ethically and responsibly. Approached wisely, we may be able to utilize AI as a sophisticated amplifier of human ingenuity. This utilization can help promote innovation far beyond what we envision today across medicine, industry, and the very nature of what it means to create in the 21st century.

Generative AI in Creative Fields

Human creation, whether painting, composing, or writing, has always carried an element of surprise. When an idea formed in the artistic mind is being painted on a canvas, there is a strong feeling associated with giving the idea a tangible shape. Generative AI introduces a new collaborator into that process, allowing for a dialogue between artist and algorithm.

Let's consider another possible scenario: a visual artist inspired by the movement of clouds at dusk. The artist can feed that abstract concept into a generative AI system trained on landscapes. The artist might type a prompt such as "majestic skyscape, dappled with golden light, moving like smoke," and the AI can generate unique images that match the initial prompt. Is this truly a collaboration if the human merely refines or curates the AI's output? Does art lie in the idea or the execution? These questions will redefine how we value and approach artistic practice in the decades to come.

Perhaps a more fascinating idea to consider is how generative AI can transform the artistic process. Music producers might experiment with tools that take a whistled melody and create the entire song. They could ask the AI to create layers of instrumentation, from classical orchestration to synth-heavy electronics. Authors writing about a vibrant marketplace could have an AI generate snippets of overheard conversations, each with distinct voices and tones, and inspire new ideas they might not have considered on their own.

Generative AI is not only a tool for quickly making artwork. Visual artists with limited technical skills can better design their ideas without mastering specialized software. Similarly, writers can evoke complex sensory imagery using AI, improving their writing without losing ownership of the story. Generative AI can help many different kinds of artists create their work quickly and efficiently and may even inspire them.

Generative AI, however, does not automatically confer creativity onto its user. The true innovation will come from humans understanding the strengths and limitations of these systems. Some artists will manipulate, hack, and subvert what the generative models produce. The human eye and discernment give AI-generated artworks the power to speak to our shared experience and challenge our perceptions.

It is crucial, too, to consider the ethical questions around intellectual property. How do we determine the difference between AI influence and direct AI production if a painting utilizes generative content? We urgently need nuanced legal frameworks to help with emerging issues raised by the ability of machines to emulate artistic skill.

It is important to consider the challenges of generative AI now, before we use it more widely. We must engage with these types of AI tools strategically, thoughtfully, and critically as society redefines what it means to be an artist.

The Future of Employment in the Age of Generative AI

Throughout history, new technologies have brought anxieties about job displacements. This anxiety happened with the mechanization of agriculture, again with industrial automation, and once more with the rise of computers. Today, generative AI has caused new concerns about the future of work and human employment. However, as is often the case, a deeper discussion reveals potential beyond straightforward narratives of disruption.

Generative AI undoubtedly can automate many tasks previously performed by humans. For example, consider customer service representatives: AI chatbots may soon address the vast majority of routine queries, potentially displacing those currently working in those positions. Similarly, copywriters producing basic articles or product descriptions could compete with AI engines proficient in writing persuasive copy. This highlights the immediate need for proactive upskilling and retraining programs to mitigate the impact on these types of workers.

Yet, it is equally true that generative AI will bring about the birth of entirely new professions. There will be "prompt engineers" specializing in coaxing the best output from generative models, AI ethicists navigating the technology's complex legal and societal dimensions, and specialists who ensure datasets used to train AI remain bias-free to the greatest extent possible. These nascent roles signify a change in the skills economy, not an inherent devaluation of the human workforce.

Perhaps generative AI's more subtle transformative effect will be how it fundamentally alters existing roles. Accountants could work alongside AI tools that identify discrepancies, anomalies, and patterns within vast financial datasets at incredible speed. Architects may find generative AI a limitless inspiration engine, suggesting bold spatial possibilities and saving countless hours of technical sketching. Professionals will need to move beyond being skilled operators of specific tools to strategists capable of orchestrating, refining, and curating the output of generative AI systems.

It is important to acknowledge that there will inevitably be disruption as these emerging AI systems get more sophisticated. History proves that technological advancements bring both progress and societal changes. Lawmakers, businesses, and educators must foster an environment where these changes do not simply happen haphazardly. We must proactively discuss solutions like retraining programs targeted at vulnerable skill sets, revised labor laws designed for the realities of human-AI collaboration, and exploring concepts like universal basic income should widespread displacement materialize.

The future of work with generative AI requires a sober assessment of its potential perils and an unwavering belief in human ingenuity. The goal needs to be a thriving human-AI symbiosis, where automation allows us to focus on complex cognitive tasks, higher-order problem-solving, and those crucial areas where genuine human empathy remains essential.

Policy Implications and Regulatory Challenges

Let us consider the current regulatory frameworks we have around emerging technologies. They were generally developed after those technologies matured through real-world use. The Internet is a fantastic case study; regulations surrounding content, consumer protection, and digital property rights took years and often heated debates to catch up with the speed of digital transformation. With generative AI, we find ourselves in an uncanny position, facing monumental questions concerning societal impacts, accountability, and even the core philosophical boundaries between human and machine creativity well before these systems attain their fullest potential.

One fundamental issue of regulating generative AI is the concept of *black box* systems, computer programs that are not transparent in how they operate or achieve their results. These systems have been mentioned a few times in the previous chapters of this book. Large generative models process enormous quantities of data to arrive at results that are difficult for even their creators to explain. When these systems create content flagged as potentially harmful or generate responses that appear biased or discriminatory, pinpointing the precise source of the error within the billions of connections they have formed poses a regulatory challenge.

Furthermore, the very act of generating synthetic content that appears indistinguishable from the human-authored original is a problem. While such technology allows for creative possibilities, it equally carries the threat of deepfakes or intentionally deceptive media at scale. How do we verify the integrity of content when generative AI makes that line difficult to identify? What are the regulatory boundaries in creative fields, such as credit, compensation, and ensuring that imitators do not harm the original artists?

There is also the crucial matter of who bears responsibility if an AI-generated product, content, or decision causes tangible harm. Is it the developer, the datasets used to train the model, or the individual deploying the technology for unforeseen purposes? Liability in regard to advanced machine learning is a challenging determination to make. Without adequate safeguards, we risk stifling innovation while simultaneously being unprepared against malicious applications.

It is neither advisable nor even feasible to aim for a heavy-handed approach that does not permit generative AI innovation. Failing to create reasonable policy about AI risks serious implications as this technology embeds itself deeply in fields ranging from media, marketing, art, and even law. Policymakers must embrace a mindset of agile adaptation and collaborative development with technology experts, legal scholars, and ethicists working collectively to craft workable regulatory frameworks for AI.

The public must be aware that AI rules and laws are not only for specialized committees to consider. The regulations for emerging technologies impact everyone, influencing what information we can trust online, how our future jobs are structured, and how we manage AI's influence on creativity.

PUBLIC PERCEPTION AND SOCIETAL IMPACT

The rise of generative AI presents us with a curious paradox. For example, people are pleased with chatbots capable of intelligent conversation, seeing realistic photo edits at the click of a button, and experimenting with AI-generated art. AI systems are delivering incredible services without the need of mastery of specific creative tools or disciplines.

At the same time, public perception around generative AI contains anxieties, some justified and some based on speculation and media hype. These concerns range from existential worries about jobs being replaced by machines to legitimate issues regarding AI bias and a mistrust of AI-manufactured content. Part of the anxiety is due to these systems and their inherent opacity that most consumers cannot understand.

This complexity in public perception demands an open, multifaceted conversation. One way to approach distrust in AI systems is with public education. Explaining in everyday language how generative AI functions, both its strengths and limitations, allows for improved public awareness, not one purely governed by hype or fear.

Education will allow for the improved and informed formation of public opinion rather than a public opinion formed from exaggerated news reports. This public education extends to discussions about how generative AI changes notions of creative authorship, how to prevent discriminatory or misleading use, and, perhaps most importantly, the skills to develop to thrive in this rapidly evolving landscape.

Another significant area to explore is how generative AI reshapes societal expectations, especially considering access and equality. The potential of AI systems to address skill gaps in fields such as medicine or architectural design could have real-world social benefits. Yet, if their access is concentrated in the hands of already privileged populations or corporations, these tools risk solidifying existing inequalities rather than being forces of true fairness. Questions surrounding how to ensure equitable access to such transformative technologies deserve a place in wider debate and policy dialogue.

There is also a fascinating question of identity as generative AI weaves is utilized in more online interactions. When the online profile and digital presence of the average person utilize generative avatars, curated personas, or content partially shaped by AI tools, are these still authentic representations of the user? What are the philosophical and societal ramifications of the online identity increasingly becoming AI-mediated?

The public perception of generative AI should not only include education but proactive, collaborative discussions that anticipate issues before they become a problem. Managing the public perception of generative should not be about fostering technological utopianism or fearmongering. Instead, it should be about a society-wide reckoning of generative AI as a tool that is being incorporated into our daily experiences.

Case Study: Real-World Implementation of a Generative AI System

Let's consider a practical example that illustrates the challenges and incredible power of implementing generative AI within a real-world context. Jasper.ai is a company utilizing these models to transform how content is written. Their platform specializes in creating content for various types of material, such as social media posts, marketing copy, and long-form blog articles. The core functionality of the company's AI is an improvement over systems that generate generic content.

Jasper.ai relies on massive datasets of existing content combined with user input parameters. For example, let's say a small business needs marketing copy for its organic skincare Web site. It can supply basic information like keywords, desired tone (concise and informative, perhaps), and target audience characteristics. Using generative AI, the platform generates original product descriptions, introductory paragraphs, and a variety of headlines for A/B testing. Jaspar.ai speeds up the time-consuming writing tasks that can trouble small businesses with limited marketing resources.

The value from this system is derived from human collaboration. The generated content does not eliminate the need for a skilled copywriter. Instead, it provides starting material with various options. The advantage of using this AI emerges when a human is used to edit, curate, and fine-tune the AI-generated copy. Further, as the system incorporates feedback from a specific business on which outputs perform best, Jasper.ai continuously learns and improves its results within that individual use case.

This use case illustrates the importance of context and human judgement. Generative AI may reduce some types of work, but the responsibility is on the human user to shape the work according to business objectives, refine for brand voice, and ensure accuracy. This case study also reveals potential scale problems when content generation is heavily automated; ensuring the absence of factual errors becomes much more complex than on a smaller scale.

There is an intriguing lesson for other enterprises considering the integration of generative AI, as well. The most successful implementations may leverage these tools to address particular problems or augment workflows. Rather than focusing on the replacement of human creativity, targeting bottlenecks and time-consuming, repetitive tasks allows for measurable returns on investment, the refinement of outputs, and the ability to develop best practices around the human-AI workflow.

As generative models improve, AI like Jasper.ai represents only the start. Additional AI tools will be improved to augment what we are capable of, especially within time and resource constraints, and their significance will only increase. Human workers should not fear these emerging technologies but acquire the acumen to utilize them effectively.

Predictions and Future Outlook

The potential for generative AI tools is considerable. It is wise to remember that technology's true impact often reveals itself retrospectively, just as few people predicted how dramatically social media or smartphones would alter our daily lives. We are likely going to underestimate the transformations yet to come with generative AI. Let's outline several distinct areas where developments hold fascinating, sometimes disruptive, potential and discuss their implications.

Personalized experiences

Let us consider another possible future scenario: a world where virtually every online interaction is tailored to the individual user. Ads are not generically designed for a large audience; instead, they dynamically target an individual's preferences and browsing history. E-commerce sites can provide better recommendations, presenting products in the user's visual styles and descriptions that match their personal vocabulary. While the personalization of marketing is likely of considerable interest to business, it raises questions about privacy. How do we balance convenience and users' control of their data? Can we avoid problems created by the AI's ability to personalize materials?

Transforming scientific inquiry

Generative AI's pattern-detecting ability is already utilized in medicine and material science. The next area for its use may be scientific breakthroughs supported by human research as well as by AI systems, which can suggest unexplored directions or identify novel connections hidden in the data. We might discover cures faster, optimize renewable energy systems, and engage in scientific advancement at an unprecedented scale, provided these tools are employed with the utmost scientific rigor.

Generative content usage and creativity

We are witnessing the initial wave of AI-powered image, video, and music generation. Once these tools become commonplace and are less computationally expensive, more synthetic content will likely follow. While concerns about provenance, copyright, and deepfakes pose immediate challenges, AI can support independent artists with limited budgets to make films or record orchestral scores. AI can affect what it means to be a creator and how we attribute content.

Evolving skillsets and "metacreativity"

Technical proficiency with specific digital tools may become less important because of generative AI. Instead, skillsets like *prompt engineering* (writing AI prompts to extract optimal results from these models) and being able to refine synthetically produced content may become much more important. Generative AI may create a new type of worker, *metacreatives*, who are individuals adept at curating, orchestrating, and refining what the machines produce while leveraging human insight that no algorithm can fully replicate.

Generative AI in public discourse

While much attention has focused on the creative and consumer aspects of AI use, consider the impact when generative AI enters the arena of public discourse. What happens when personalized propaganda campaigns become sophisticated? How do we maintain truth and trust in information when synthetic materials become more difficult to discern? Policymakers and technologists must address the questions about political influence created through generative misinformation campaigns.

This is not an exhaustive list of possible future considerations for generative AI. One important idea is the same in these predictions: The most pivotal questions are less about purely technological advances and more about the social structures, regulatory frameworks, and human capabilities we must cultivate if we intend to beneficially use these AI tools. Our task is to boldly define how humanity benefits from these tools without allowing their misuse.

APPENDIX

REFERENCES

- Beth Kindig, a Technology Analyst published in Beth. *Technology.* 2020. [Last accessed on 30 Mar 2023]. Available from: https://wwwforbescom/sites/bethkindig/2020/01/31/5-soon-t0-be-trends-in-artificial-intelligence-and-deep-learning/
- *Bostrom Nick: Superintelligence: paths, dangers, strategies.* Keith Mansfield: Oxford University Press; 2014.
- Delcker J. *Politico Europe's artificial intelligence correspondent told DW News in DW News on Black Box of Artificial Intelligence.* 2018. [Last accessed on 2013 May 18]. Available from https://mdwcom/en/can-ai-be-free-of-bias/a-43910804 .
- Dina B. *"Microsoft develops AI to help cancer doctors find the right treatment" in Bloomberg News.* 2016
- European Commission on Ethical Guidelines for Trustworthy AI. *The High-Level Expert Group on AI presented this guideline which stated three requirements: lawful, ethical and robust*
- Jacob R. Thinking machines: The search for artificial intelligence. *Distillations.* 2016;2:14–23.
- Voigt, P., & Von Dem Bussche, A. (2017). The EU General Data Protection Regulation (GDPR). In Springer eBooks. https://doi.org/10.1007/978-3-319-57959-7
- Jerry K. *Artificial Intelligence – what everyone needs to know.* New York: Oxford University Press; 2016.
- Joseph W. *Computer Power and Human Reason from Judgement to Calculation.* San Francisco: W H Freeman Publishing; 1976.
- Kaplan A, Haenlein M. Siri, Siri, in my hand: Who's the fairest in the land? On the interpretations, illustrations, and implications of artificial intelligence. *Business Horizons.* 2019; 62:15–25.
- Meera S. *Are autonomous Robots your next surgeons CNN Cable News Network.* 2016

- Nature News, 24 January 2020. *The battle for ethical AI at the world's biggest machine-learning conference by Elizabeth Gibney.* [Last accessed on 2023 Sept 13]. Available from: https://www.nature.com/articles/d41586-020-00160-y.
- Nick B, Yudkowsky E. The Ethics of Artificial Intelligence. In: Keith Frankish, William Ramsey., editors. *Cambridge Handbook of Artificial Intelligence.* New York: Cambridge University Press; 2014.
- Nils N. *Artificial Intelligence: A New Synthesis.* Morgan Kaufmann; 1998.
- Nilsson JN. *Principles of artificial intelligence.* Palo California: Morgan Kaufmann Publishers; 1980.
- Prof Stephen Hawking, one of Britain's pre-eminent scientists, has said that efforts to create thinking machines pose a threat to our very existence. *Interview on BBC on Dec 2, 2014.* Noted by Rory CellanJones.
- Quoted from Nathan Strout: The Intelligence Community is developing its own AI ethics on Artificial Intelligence Newsletter. 2020. [Last accessed on 2023 May 21]. Available from: https://wwwc4isrnetcom/artificial-intelligence/2020/03/06/the-intelligence-community-is-developing-its-own-ai-ethics/
- Roger C. Schank.Where's the AI. *AI Magazine.* 1991;12:38.
- Rory CJ. *Stephen Hawking warns artificial intelligence could end mankind BBC News Wikipedia, the Free Encyclopedia on Artificial Intelligence.* 2014. [Last accessed on 2013 Jun 23]. Available from: https://enwikipediaorg/wiki/Artifical_Intelligence .
- Russell SJ, Norvig P. *Artificial Intelligence: A Modern Approach.* Upper Saddle River, New Jersey: Prentice Hall; 2009.
- Scoping study on the emerging use of Artificial Intelligence (AI) and robotics in social care published by Skills for Care. [Last accessed on 2019 Aug 15]. Available from: wwwskillsforcareorguk .
- Von der Leyen. the President of European Commission unveiled EU's plans to regulate AI on Feb 19. 2020. [Last accessed on 2023 Jul 08]. at www.dw.com/en/european-union-unveils-plan-to-regulate-ai/a-52429426 .
- Charles S. Elliott, "JACME2T: An industry - academic consortia to enhance continuing engineering education", FIE Conference, 1998.
- Khalid Isa, Shamsul Mohamad and Zarina Tukiran, "Development of INPLANS: An analysis on students' performance using neuro-fuzzy", Symposium on Information Technology, vol 3, pp. 1–7, 2008.
- Carlos Márquez-Vera, Cristóbal Romero Morales and Sebastián Ventura Soto, "Predicting school failure and dropout by using data mining techniques", IEEERITA, vol. 8, pp. 7–14, 2013.
- Usamah bin Mat, Norlida Buniyamin, Pauziah Mohd Arsad and Rosni Abu Kassim, "An overview of using academic analytics to predict and improve students' achievement: A proposed proactive intelligent intervention", IEEE Conference on Engineering Education (ICEED), pp. 126–130, 2013.
- Hsu-Chen Cheng and Wen-Wei Liao, "Establishing a lifelong learning environment using IoT and learning analytics", ICACT, pp. 1178–1183, 2012.

- Carlotta Schatten, Martin Wistuba, Lars Schmidt Thieme Sergio and Gutierr´ez-Santos, "Minimal invasive integration of learning analytics services in intelligent tutoring systems", ICALT, pp. 746–748, 2014.
- T. Rohloff, S. Oldag, J. Renz and C. Meinel, "Utilizing web analytics in the context of learning analytics for large-scale online learning," 2019 IEEE Global Engineering Education Conference (EDUCON), Dubai, United Arab Emirates, pp. 296–305, 2019.
- Liaqat Ali, Marek Hatala, Dragan Gašević and Jelena Jovanović, "A qualitative evaluation of evolution of a learning analytics tool", Computers & Education, Vol. 58, No. 1, pp. 470–489, 2012.
- Anna Lea Dyckhoff, Dennis Zielke, Mareike Bültmann, Mohamed Amine Chatti and Ulrik Schroeder, "Design and implementation of a learning analytics toolkit for teachers", Journal of Educational Technology & Society, Vol. 15, No. 3, pp. 58–76, 2012.
- S. Dawson, A. Bakharia and E. Heathcote, "SNAPP: Realising the affordances of real-time SNA within networked learning environments", Learning, pp. 125–133, 2010.
- Alyssa Friend Wise, Yuting Zhao and Simone Nicole Hausknecht, "Learning analysis for online discussions: A pedagogical model for intervention with embedded and extracted analytics", Proceedings of the Third International Conference on Learning Analytics and Knowledge, pp. 48–56, 2013.
- A. del Blanco, A. Serrano, M. Freire, I. Martínez-Ortiz and B. Fernández-Manjón, "E-Learning standards and learning analytics. Can data collection be improved by using standard data models", Global Engineering Education Conference (EDUCON) IEEE, pp. 1255–1261, 2013.
- Hendrik Drachsler and Wolfgang Greller, "The pulse of learning analytics understanding ings and expectations from the stakeholders", Proceedings of the 2nd international conference on learning analytics and knowledge, pp. 120–129, 2012.
- Leah P. Macfadyen and Shane Dawson, "Numbers are not enough. Why e-learning analytics failed to inform an institutional strategic plan", Journal of Educational Technology & Society, Vol. 15, No. 3, pp. 149–163, 2012.
- Doug Clow, "MOOCs and the funnel of participation", Proceedings of the 3rd International Conference on Learning Analytics and Knowledge, pp. 185–189, 2013.
- Bohong Yang, Zeping Yao, Hong Lu, Yaqian Zhou and Jinkai Xu, "In-classroom learning analytics based on student behavior, topic and teaching characteristic mining", Pattern Recognition Letters, Vol. 129, 2019.
- Goodfellow, I., Pouget-Abadie, J., Mirza, M., Xu, B., Warde-Farley, D., Ozair, S., Courville, A., Bengio, J. (2014). Generative Adversarial Networks. in Proceedings of the International Conference on Neural Information Processing Systems (NIPS), pp. 2672—2680.
- Hu, W., Tan, Y., Generating Adversarial Malware Examples for Black-Box Attacks Based on GAN, [Online]. Available at: https://arxiv.org/pdf/1702.05983.pdf.

Guo, S., Zhao, J., Li, X., Duan, J., Mu, D., Jing, X. (2021). A black-box attack method against machine-learning-based anomaly network flow detection models, Security and Communication Networks, 2021(5578335), 13. doi: https://doi.org/10.1155/2021/5578335.

Shahpasand, M., Hamey, L., Vatsalan, D., Xue, M. (2019). Adversarial attacks on mobile malware detection. In Proceedings of 2019 IEEE 1st International Workshop on Artificial Intelligence for Mobile (AI4Mobile), pp. 17–20. doi: https://doi.org/10.1109/AI4Mobile.2019. 8672711.

Kargaard, J., Drange, T., Kor, A., Twafik, H., Butterfield, E. (2018). Defending IT systems against intelligent malware. In Proceedings of 2018 IEEE 9th International Conference on Dependable Systems, Services and Technologies (DESSERT) pp. 411–417. doi: https://doi. Org/10.1109/ DESSERT.2018.8409169.

Taheri, R., Shojafar, M., Alazab, M., & Tafazolli, R. (2021). Fed-IIoT: a robust federated malware detection architecture in industrial IoT. Proceedings of IEEE Transactions on Industrial Informatics, 17(12), 8442–8452. https://doi. org/10.1109/TII.2020.3043458

Kim, J., Bu, S., & Cho, S. (2017). Malware detection using deep transferred generative adversarial networks. Proceedings of ICONIP, Part I, LNCS, 10634, 556–564. https://doi.org/10. 1007/978-3-319-70087-8_58.

Kim, J., Bu, S., & Cho, S. (2018). Zero-day malware detection using transferred generative adversarial networks based on deep autoencoders. Information Sciences, 460–461, 83–102. https://doi.org/10.1016/j. ins.2018.04.092

Hitaj, B., Gasti, P., Ateniese, G., Perez-Cruz, F. (2019). PassGAN: A Deep Learning Approach for Password Guessing, 2019, [Online]. Available at: https://arxiv.org/abs/1709.00440.

Struik, O., Kondratenko, Y. (2021). Generative adversarial neural networks and deep learning: successful cases and advanced approaches, International Journal of Computing 339—349. doi: https://doi.org/10.47839/ ijc.20.3.2278.

W. Knight, The Defense Department has produced the first tools for catching deepfakes, 2018, [Online]. Available at: https://www.technologyreview. com/2018/08/07/66640/the-defense-dep   artment-has-produced-the-first-tools-for-catching-deepfakes/.

Nguyen, T. T., Nguyen, Q. V. H., Nguyen, D. T., Nguyen, D. T., Huynh-The, T., Nahavandi, S., Nguyen, T. T., Pham, Q.-V., Nguyen, C. M. (2022). Deep Learning for Deepfakes Creation and Detection: A Survey, [Online]. Available at: https://arxiv.org/abs/1909.11573.

Agarwal, S., Varshney, L. R. (2019). Limits of Deepfake Detection: A Robust Estimation Viewpoint [Online]. Available at: https://arxiv.org/ abs/1905.03493.

Maurer, U. M. (2000). Authentication theory and hypothesis testing. Proceedings of IEEE Transactions on Information Theory, 46(4), 1350–1356. https://doi.org/10.1109/18.850674

ABOUT THE AUTHOR

With a career spanning almost three decades, Enamul Haque has been a keen observer and active participant in the whirlwind evolution of the modern technology landscape. From early coding projects whispered into life on university mainframes to today's cutting-edge AI-driven transformations, he's always seemed to have a knack for being at the heart of where innovation happens. His path, marked by collaborations with global giants like Microsoft, Nokia, HCL Teach Capgemini and Wipro, to impactful roles driving change within the United Nations – speaks to a restless spirit of exploration and a dedication to utilizing technology as a force for good.

Perhaps best described as a "data whisperer" (he has an ear for the patterns hiding in vast information sets), Enamul is recognized for his expertise in areas like Intelligent Process Automation, service integration, and the strategic dance of digital transformation. Fortune 500 companies lean on his ability to decipher the ever-shifting technological puzzle, ensuring success and genuine business evolution. Enamul's work isn't merely about implementing the newest shiny tool but creating harmony between systems, data, and the human teams they empower.

With an unquenchable thirst for knowledge, Enamul's credentials reflect ongoing exploration, whether in formal settings like Harvard Business School and Helsinki's renowned machine learning program or the hands-on understanding born from his prolific output as an author and researcher. From IT service management to the vast oceans of data that fuel AI, few topics escape his fascination and drive to contribute. This blend of practical application and insightful writing has positioned him as a valued speaker, with the University of Coventry proudly hosting his lectures to students eager for real-world wisdom.

But Enamul isn't confined to corporate boardrooms or academic campuses. His work spans continents, and his contributions to humanitarian missions speak of a belief in making technology serve real people those facing crisis and those striving for progress in emerging economies. Whether sharing his insights with young professionals looking for their own path or lending his skills to global non-profit initiatives, Enamul embodies his values by using his position in the ever-advancing field of IT to create a positive impact wherever possible.

Beyond boardrooms and vast technical projects, Enamul brings the transformative power of tech to eager minds as the author of acclaimed technical guides. His best-selling "The Ultimate Modern Guide to Cloud Computing" is a testament to his clarity. The Computer Science Department at the University of West England ranked him a top resource. Enamul tackles some of the field's hottest topics, with additional best-selling works on AI, IoT, Data Science, and Digital Transformation serving as valuable resources for both aspiring and experienced tech professionals. This ability to distill complex information into easily grasped knowledge proves a potent asset, his expertise flows out from codebases into the classroom, shaping the next wave of technical innovators.

*I*NDEX

www.ingramcontent.com/pod-product-compliance
Lightning Source LLC
Chambersburg PA
CBHW071420050326
40689CB00010B/1909